ATLAS OF THE BRYOPHYTES OF BRITAIN AND IRELAND

1. LIVERWORTS
(HEPATICAE AND ANTHOCEROTAE)

ATLAS OF THE BRYOPHYTES
OF
BRITAIN AND IRELAND

VOLUME 1. LIVERWORTS
(Hepaticae and Anthocerotae)

EDITED BY

M. O. HILL
Institute of Terrestrial Ecology

C. D. PRESTON
Institute of Terrestrial Ecology

AND

A. J. E. SMITH
University College of North Wales

HARLEY
BOOKS

1991

Published by Harley Books
(B. H. & A. Harley Ltd.)
Martins, Great Horkesley,
Colchester, Essex, CO6 4AH, England
for
the British Bryological Society
c/o Department of Botany,
National Museum of Wales,
Cardiff, CF1 3NF

with the support of
The Natural Environment Research Council
and
The Nature Conservancy Council

Designed by Geoff Green
Text set in Ehrhardt by Saxon Printing Ltd, Derby
Printed in Great Britain by St Edmundsbury Press Ltd
Bury St Edmunds, Suffolk

British Library Cataloguing-in-Publication Data
1. Mosses
I. Hill, M. O. (Mark O.) *1945*–   II. Preston C. D. (Christopher David) *1955*–
III. Smith, A. J. E. (Anthony John Edwin) *1935*–
Atlas of the Bryophytes of Britain and Ireland
Volume 1. *Liverworts (Hepaticae and Anthocerotae)*
588.33

ISBN *0 946589 29 1*

CONTENTS

FOREWORD

Dot-distribution maps based on National Grids have become a standard method for reporting the distribution of plant and animal species in Britain and Ireland. Maps for selected bryophyte species have been published by the British Bryological Society, but now a full set of maps covering nearly 1000 taxa of mosses and liverworts is to join the growing series of Atlases of our native flora and fauna. This volume on liverworts is the first in a series of three; the two subsequent volumes will be devoted to mosses.

Distribution maps are valuable in several respects. They are essential to the study of biogeography, and allow species to be assigned to categories of distribution pattern. On the global or continental scale there are elements such as Arctic-Alpine, Atlantic, Mediterranean, and so on. Not surprisingly, the British-Irish ranges are often a small-scale reflection of these wider geographical patterns. There is probably still wider interest in relating distribution of species to the controlling influences of environmental factors, especially climate and geology/soil. Close correlations between species' distribution limits and parameters of temperature, precipitation and other factors can provide some extremely suggestive insights about climatic control of distribution, though these need to be examined in depth and, preferably, subject to experimental testing under controlled conditions. The ecological value of species' distribution maps is revealed by comparison with appropriate environmental factors. A further extension of this insight is in using maps to judge which blank areas would be worth searching for new localities of a species.

The third main value of species' maps is of particular importance to nature conservation, by showing national abundance and hence identifying the rare and endangered species, and other more widespread taxa undergoing appreciable national or regional change. While the comparison of pre- and post-1950 records can help to identify declining or increasing bryophyte species, great care is needed in using the maps as a monitoring tool. In many cases, an open circle on a map may merely indicate that nobody has been back to search since 1950, or that the particular species has proved elusive. Yet, when a consistent failure to re-confirm old records matches a pattern of relevant habitat loss or pervasive environmental change, the maps can be a convincing demonstration of adverse human impact on our wild flora and fauna. It is a general criticism of dot distribution maps that, especially for the more obscure and taxonomically difficult groups, they may be more a measure of the distribution of recorders and the intensity of recording effort than of the occurrence of the species concerned.

The failure of dot-distribution mapping to measure local abundance is more a reflection of the scale of mapping than of defects in the technique. The larger the scale, the more truly does the method reflect abundance. For some species' groups and for certain areas, 'tetrad' mapping is steadily giving a finer resolution for distribution mapping. For most bryophytes, we are not yet at this stage, and only for a few species is there any supplementary information on population size. Nevertheless, the publication of the Atlas of Bryophytes represents a great achievement considering the relatively small number of field bryologists in the British Isles. As with maps for other species groups, it is a tribute especially to the industry and enthusiasm of both professional and amateur workers, whose efforts were previously reported in the annual lists of new vice-county records for Britain and Ireland, and in the Census Catalogues. It will be of great interest to bryologists more widely, in expanding our knowledge of one of the richest moss and liverwort floras in Europe, and one of the internationally important biological features of Britain and Ireland.

Cambridge
December, 1989

D. A. RATCLIFFE

PREFACE

There has been keen interest in the distribution of British bryophytes for over ninety years, ever since the foundation of the Moss Exchange Club in 1896. With an increase in field bryology and the rejuvenation of the British Bryological Society after the Second World War, and stimulated by the activity of the Botanical Society of the British Isles in recording the distribution of vascular plants, the BBS Distribution Maps Scheme was launched in 1960.

Apart from simply recording the presence or absence of mosses and liverworts in the 10-km squares of the British and Irish National Grids, there were a number of underlying objectives, either explicit or implicit. There is no better way of understanding bryophytes than by recording the occurrence of species in a variety of habitats. There was a growing awareness of the effects upon our flora of industrialization, urbanization and pollution and the feeling of a need to monitor change in frequency consequent upon these. The Mapping Scheme would, in addition to field recording, draw together information from many disparate sources in the literature and herbaria. On a more parochial level, it would provide a purpose for recording on BBS field meetings and would indicate to the world the active state of British bryology.

Whilst the data assembled here are patently incomplete it is hoped the maps will provide a reasonably accurate picture of the distribution of mosses and liverworts in the British Isles and stimulate further field activity, both in recording and in other aspects of bryology.

Bangor   A. J .E. SMITH
December, 1989

9

ACKNOWLEDGEMENTS

This atlas is based on the fieldwork of members of the British Bryological Society; we take this opportunity to express our appreciation of the dedication and enthusiasm with which they have undertaken this task. The financial support of the Natural Environment Research Council and the Nature Conservancy Council has enabled us to turn the results of this work into a published atlas. We thank all those who have helped to prepare the first volume for publication, by data processing, checking draft maps and writing notes on the species. Detailed acknowledgements, including a list of recorders, will be published in the final volume. Publication of the atlas has been assisted by grants from the Linnean Society of London and the Royal Society.

HISTORY OF BRYOPHYTE RECORDING
IN THE BRITISH ISLES

C. D. PRESTON

THE BEGINNING OF BRYOLOGY

Knowledge of the taxonomy and distribution of bryophytes has always lagged behind that of vascular plants, for the simple reason that they are so much smaller and less conspicuous. The herbalists had little exact knowledge of cryptogams; G. A. M. Scott (1987) comments that 'identification was chaotic and the surprising thing is not how much error there was, but how anyone got any plants right at all'. Only a few species were recorded by the earliest modern botanists. John Ray, for example, listed just six bryophytes in his pioneer flora of Cambridgeshire, the *Catalogus Plantarum circa Cantabrigiam nascentium* (1660). One of these, *Adiantum aureum majus*, recorded 'on Hinton Moor in the watery places', is believed to be *Polytrichum commune* (Proctor, 1956). The identity of the remainder can only be guessed at and one, *Muscus ex cranio humano*, 'the Mosse on a dead mans skull', is probably completely mythical (Scott, 1988). A further two species were added by Ray (1663) in the first appendix to his *Catalogus*, including *Muscus triangularis aquaticus*, clearly *Fontinalis antipyretica*, 'in the river beyond Stretham ferry'.

Ray's subsequent publications reveal the increasing knowledge of bryophytes. He recognized over 80 species in his last work on the British flora, the second edition of *Synopsis methodica Stirpium Britannicarum* (1696). Some of these were the result of Ray's own exploration of the area round his home in Black Notley, Essex, including *Riccia glauca 'in horto nostro inveni'* (discovered in our garden) and *Lunularia cruciata*, 'found on moist and shady banks in Essex, and doubtless to be found in the like places elsewhere'. Many more species were based on material sent to Ray by correspondents such as Edward Lluyd, whose record of *Racomitrium lanuginosum*, *'vitium graminis in montosis Cambriae'* (a blemish of grassland in mountainous Wales), was published by Ray under the cumbersome polynomial *Muscus terrestris vulgari similis lanuginosus*.

After Ray's death a new edition of the *Synopsis* (1724) was prepared by Johann Jakob Dillenius. Dillenius, German by birth, was persuaded to emigrate to England in 1721 by the botanist and botanical patron William Sherard. He had been recommended to Sherard as 'a person very curious in mushrooms and mosses' (Turner, 1835, p. 146) and he was able to add over 60 bryophyte species to the *Synopsis*. Many of these resulted from fieldwork in the Home Counties, others were based on specimens sent from elsewhere by correspondents of Sherard and Dillenius. The *Synopsis* remained the standard British flora for nearly 40 years, but even before its publication, Dillenius was working on a much more ambitious bryological work (Turner, 1835, p.211). This was eventually published in 1741 as *Historia Muscorum*, a title perhaps modelled (consciously or unconsciously) on Ray's massive *Historia Plantarum* (1686–1704). *Historia Muscorum* includes an illustrated account of the bryophytes then known to science, and is a book of such significance that its publication is often thought to mark the beginning of systematic bryology. It incorporates many of Dillenius' own British records, including those made with Samuel Brewer on their exploration of North Wales in 1726, when they worked 'country ... very productive for Mosses and Bogg plants' (Druce & Vines, 1907). Some of the characteristic northern and western bryophytes were discovered on this journey (Richards, 1979), including *Anthelia julacea* from Cader Idris, Snowdon and Glyder, *Bryum alpinum*, so conspicuous near Llanberis that Dillenius was astonished that Lluyd had overlooked it, and, most notable of all, *Scapania ornithopodioides*, detected growing amongst mosses on Snowdon, where it still occurs.

The years following the publication of *Historia Muscorum* were rather barren for British bryology, as indeed they were for British natural history as a whole (Allen, 1976). Linnaeus' *Species Plantarum* (1753) introduced binomial nomenclature, but the author was becoming both mentally and physically exhausted as he approached the completion of his work, and his accounts of bryophytes are based primarily on *Historia Muscorum* (Isoviita, 1970). The Linnaean binomials were taken up by W. Hudson in his *Flora Anglica* (1762) and by J. Lightfoot in *Flora Scotica* (1777). Lightfoot's book broke with tradition in describing species in English rather than Latin, 'for the use of my countrymen, who will understand it never the worse for being in their own tongue'.

EXPLORATION OF THE BRYOPHYTE FLORA OF THE BRITISH ISLES, 1785–1855

One of the first signs of increasing interest in bryology was the *Fasciculus Plantarum Cryptogamicarum Britanniae* (1785), written by the Covent Garden nurseryman James Dickson. Numerous species were added to the British flora in this and the three subsequent fascicles (Dickson, 1790–1801), many from localities in the London area (including Croydon, Enfield Chase, Hampstead Heath and Muswell Hill), others the result of Dickson's fieldwork in Scotland. Although some of the Scottish species are unlocalized, other records show that Dickson visited Ben Lawers, Ben Lomond and Ben Nevis.

Dickson's work is less well known that it would otherwise be because his first three volumes appeared before Johannes Hedwig's posthumously published *Species Muscorum Frondosorum* (1801). This thorough revision of the known mosses, based on detailed microscopic examination, is the official starting date for the nomenclature of mosses. The title *Species Muscorum*, like that of *Historia Muscorum*, reflects that of a major vascular plant publication.

The taxonomy of the British bryophyte flora was gradually elucidated in the early 19th century in a succession of standard floras. J. E. Smith's *Flora Britannica* (1804) included mosses; Smith also provided the text for Sowerby's *English Botany* (1790–1814), in which many bryophytes were illustrated in superb colour plates. W. J. Hooker's masterly *British Jungermanniae* (1812–16) was an almost complete liverwort flora, as the genus he monographed, *Jungermannia*, then encompassed all the plants that would now be placed in the Calobryales, Jungermanniales and Metzgeriales. *Muscologia Britannica* (1818), the first flora dealing solely with British mosses, was written by Hooker in collaboration with the Anglo-Irish doctor Thomas Taylor (Sayre, 1983). An appendix covering liverworts was added to the second edition (1827). This was superseded in 1833 by a book designed by Hooker to complete two vascular plant floras, Smith's *English Flora* and Hooker's own *British Flora*. Hooker and W. Wilson contributed almost all the accounts of bryophytes to J. de C. Sowerby's *Supplements to the English Botany* (1829–66).

The exceptionally rich hepatic flora of south-west Ireland became apparent in the first decades of the 19th century. It was initially brought to light by Miss Ellen Hutchins of Ballylicky, near Bantry in Co. Cork. Miss Hutchins' fieldwork was confined to the Bantry area, she was handicapped by periods of ill-health and she died (in 1815) aged only 30. Nevertheless, she was such an acute observer that she discovered numerous hepatics new to Ireland; amongst those which were also new to science were *Adelanthus decipiens*, *Cephaloziella turneri*, *Diplophyllum obtusifolium*, *Leptoscyphus cuneifolius*, *Lophozia bantriensis* and *Radula aquilegia*. Hooker described these in *British Jungermanniae*, which began with *Jungermannia* (now *Jubula*) *hutchinsiae*, discovered by Miss Hutchins in 'gloomy caverns' near Bantry and named by Hooker in her honour. T. Taylor also discovered a number of Atlantic hepatics, including *Aphanolejeunea microscopica*, *Plagiochila punctata*, *Radula voluta* and *Scapania nimbosa*. The moss flora of Ireland was described by Dawson Turner in *Muscologiae Hibernicae Spicilegium* (1804), and both mosses and liverworts were covered by Taylor in Mackay's *Flora Hibernica* (1836). Miss Hutchins, Taylor, John Templeton of Belfast and James Drummond of Cork added many mosses to the Irish flora in this period, as did W. Wilson on a notable visit to south-west Ireland in 1829.

Interest in bryology was particularly strong in Scotland in the 1820s and 1830s. Hooker was Professor of Botany at Glasgow University from 1820 to 1841; on his appointment he wrote *Flora Scotica* (1821) for the use of his students. In Edinburgh, R. K. Greville published the *Scottish Cryptogamic Flora* (1822–28), containing plates and accompanying text after the model of *English Botany*. Greville was a founder member of the Botanical Society of Edinburgh, established in 1836 to promote 'Botanical Science in all its ramifications' (Fletcher & Brown, 1970). Many mosses new to the British Isles were discovered in the 'Scotch Alps' by Hooker, Greville and G. A. Walker Arnott, including *Blindia caespiticia*, *Campylium halleri*, *Cirriphyllum cirrosum* and *Heterocladium dimorphum*, all found on Ben Lawers. Thomas Drummond was exploring Angus at this time; his discoveries included *Grimmia unicolor* and *Stegonia latifolia* in Clova, *Timmia austriaca* near Airlie Castle and *Neckera pennata* at Fothringham. It is interesting to note that as early as 1827 Scottish botanists regarded the flora of England as almost tediously well recorded. Apologizing for describing a moss from England, *Epipterygium tozeri*, in the *Scottish Cryptogamic Flora*, Greville wrote 'Every path in England ... has been so assiduously explored that when a new plant, so high in the scale of vegetation as a Moss, is discovered, considerable interest is excited'. As William Wilson in the next few years discovered *Ephemerum sessile*, *Orthodontium gracile*, *Paludella squarrosa*, *Physcomitrium sphaericum* and *Pottia wilsonii* in the county of Cheshire, it is possible that Greville was overstating his case.

Hooker's appointment to Kew in 1841 effectively marked the end of his work on British bryophytes. It therefore fell to his friend Wilson to prepare a third edition of *Muscologia Britannica*. Wilson was an unworldly Warrington solicitor who abandoned his profession in order to pursue his interest in natural history (Robinson, 1871). His edition of the *Muscologia*, published as *Bryologia Britannica* in 1855, was actually a completely new work. Wilson's meticulous revision of the British mosses ensured that it immediately became the standard flora; Dixon (1896) was to describe it as 'the prince of bryological books'.

THE START OF SYSTEMATIC RECORDING, 1855–1914

The increasing popularity of natural history in the second half of the 19th century led to the publication of semi-popular and popular bryological books. M. J. Berkeley's *Handbook of British Mosses* (1863) and C. P. Hobkirk's *Synopsis of the British Mosses* (1873) were floras aimed at the beginner. Both acknowledged their debt to Wilson's *Bryologia Britannica*. The first truly popular work seems to have been *Twenty Lessons on British Mosses* (1846) by the Dundee umbrella-maker William Gardiner. It was followed by R. M. Stark's *Popular History of British Mosses* (1854) and F. E. Tripp's *British Mosses* (1874). Towards the end of the century *Bryologia Britannica* was replaced by R. Braithwaite's lavish *Sphagnaceae* (1880) and *British Moss-Flora* (1887–1905) and by H. N. Dixon's more modest *Student's Handbook of the British Mosses* (1896). Dixon's book, which incorporated the keys and illustrations previously published in H. G. Jameson's *Illustrated Guide to British Mosses* [1893], was both authoritative and relatively inexpensive and became the standard reference work.

There was a surprising dearth of liverwort floras in the middle of the century. B. Carrington's *British Hepaticae* (1874–75) described only 30 species before publication ceased because of poor sales and the author's declining health. The versatile author Mordecai Cubitt Cooke produced an *Easy Guide to the Study of British Hepaticae* [1865] as a supplement to the journal *Science Gossip* and a more substantial but not wholly reliable *Handbook of British Hepaticae* (1894). The situation improved in 1902, with the publication of W. H. Pearson's *Hepaticae of the British Isles* and H. W. Lett's *Species of Hepatics*. S. M. Macvicar's *Student's Handbook of British Hepatics*, a companion volume to Dixon's *Mosses*, was the standard flora from the time of its publication in 1912 until 1990.

Species continued to be added to the British flora, many from sites which were already known to be bryologically rich. *Marsupella condensata*, *Ctenidium procerrimum*, *Hypnum bambergeri*, *H. revolutum*, *Lescuraea saxicola*, *Plagiothecium piliferum* and *Scorpidium turgescens* were amongst the species discovered on Ben Lawers. *Lejeunea holtii*, *Radula carringtonii* and *R. holtii*, hepatics new to science, were described from Killarney, Co. Kerry.

The increasing popularity of bryology was reflected at the regional level. Some county floras in the 18th and early 19th centuries had covered bryophytes, including R. Relhan's *Flora Cantabrigiensis* (1785), H. Davies' *Welsh Botanology* (1813) (a flora of Anglesey), T. Power's *County of Cork* (1845) and W. Gardiner's *Flora of Forfarshire* (1848). From the 1870s the bryophytes of numerous counties, particularly in England, were surveyed in detail. Sometimes the results appeared in a volume which covered both vascular plants and bryophytes: examples include Oxfordshire (Boswell, 1886), West Yorkshire (Lees, 1888), Herefordshire (Purchas & Ley, 1889) and West Lancashire (Wheldon & Wilson, 1907). Other bryophyte floras were published as papers, usually in the *Journal of Botany* or in the transactions of a local natural history society. Counties covered in this way include Staffordshire (Bagnall, 1896), Sussex (Nicholson, 1908, 1911) and – for mosses only – Berwickshire, Northumberland and Roxburghshire (Hardy, 1868), Cheshire (Wheldon, 1898) and Northamptonshire (Dixon, 1899).

Most of the published county floras were based primarily on the fieldwork of a single bryologist. They generally followed a similar format: the habitat of a species would be described and the records then listed, often under subdivisions of the county. Their quality varies, depending on the ability of the author and the amount of time he (almost all bryologists of this period being men) was able to spend in the county. Few sank to the depths plumbed by G. C. Druce (1922), whose work was basically a transcription of Boswell's 1886 flora but incorporated a number of copying errors including the omission of a whole page of Boswell's account (Jones, 1952)! In addition to county floras, simple county lists were published, particularly in the reports of the Botanical Record Club, and innumerable lists of bryophytes from smaller areas or individual sites were published in national and local journals.

As publications on bryophytes proliferated, the need for a national synthesis of distribution data began to be felt. D. Moore (1876) summarized the distribution of Irish hepatics, citing localities for rare species but describing the distribution of common species in very general terms. F. A. Lees (1881) provided data on the distribution of bryophytes in 18 'provinces' of Britain. These provinces had been devised for vascular plants by H. C. Watson (1843). Lett (1902) also listed the distribution of liverworts in the Watsonian provinces and added data on the Irish counties; the distribution of Irish hepatics was considered in more detail by D. McArdle (1904), who listed records in twelve provinces in a work modelled on the *Cybele Hibernica* of N. Colgan & R. W. Scully (1898). However, Watson (1873–1874) had already moved beyond these coarse units and presented information for 112 'vice-counties' in Great Britain. This network was extended by Praeger (1901), who delimited 40 Irish vice-counties.

It was the Moss Exchange Club, founded in 1896, which provided the impetus needed to present the known distribution of bryophytes on a vice-county basis. One of the objects of the Club was the working up of the county distribution of bryophytes (Foster, 1979). Initially E. C. Horrell, an authoritarian and unlikeable figure in the early history of the Club, attempted to produce the necessary catalogue (Horrell, 1898). He could not complete the task, but Macvicar, who had published the vice-comital distribution of Scottish hepatics in 1904, compiled a *Census Catalogue of British Hepatics* in 1905. A companion volume for mosses was prepared by W. Ingham (1907). There was no formal mechanism for the publication of additions to the *Census Catalogues*, although specimens circulated in the annual exchange were marked with an asterisk in the reports of the Moss Exchange Club if they constituted a new vice-county record. Macvicar (1910) published a complete list of the known records of hepatics in Scotland under their vice-counties, and a second edition of the liverwort catalogue appeared soon afterwards (Ingham, 1913). The vice-county distribution of mosses in

Ireland was revised by Lett (1915). A detailed history of bryophyte recording at the vice-county scale is given by M. F. V. Corley & M. O. Hill (1981).

THE DOLDRUMS OF THE INTER-WAR YEARS

The momentum of the 19th century was not maintained after the First World War, and bryology, like other branches of natural history, declined to a low ebb. These stagnant years are well described by E. V. Watson (1985). Although the Moss Exchange Club was reorganized as the British Bryological Society in 1923, and its membership gradually increased (Richards, 1985), it appears to have had the atmosphere of a cosy club rather than that of an active scientific society. The existing activities of the Society were continued, but few new initiatives were taken. Dixon did not revise his *Student's Handbook* after 1924 and Macvicar produced the second and final edition of his in 1926, incorporating several hepatics of Mediterranean affinities added to the British flora by W. E. Nicholson. Thereafter taxonomic revisions and additions to the British flora were so few that new editions were scarcely necessary. Taxonomic attention was primarily devoted to the discussion of varieties, many of which were simply environmentally induced phenotypes. The classification of *Sphagnum*, based on the work of C. Warnstorf, developed into a system of such rococo complexity that the genus required a recorder of its own, and referees were appointed for individual sections.

The BBS maintained the system of vice-county recording: new editions of the *Census Catalogues* were prepared by J. B. Duncan (1926) for mosses and A. Wilson (1930) for hepatics. Supplements to both catalogues were issued in 1935 and all subsequent additions were published in the annual report. A *Census Catalogue of British Sphagna* was published by W. R. Sherrin (1937), followed by a second edition just after the war (Sherrin, 1946).

Few county floras were written in this period (cf. Taylor, 1954, 1955). The best was probably Wilson's *Flora of Westmorland*, published in 1938. Wilson, a septuagenarian who had been publishing papers on mosses since 1899, was a survivor of the pre-war days rather than the herald of a new dawn.

The organized activities of the BBS had to be curtailed during the Second World War. Several of the Society's stalwarts died during the war, including its leading experts on mosses and liverworts, H. N. Dixon and W. E. Nicholson.

REVIVAL AFTER 1945

The Second World War proved to be a watershed in the history of the BBS. A small group of members got together in a hotel in Borrowdale in May 1945 to discuss the revival of the Society (Perry, 1983). Their influence was apparent in the first post-war Annual Meeting, held at Appleby in September 1946. 'The view was expressed that the *Report* should be enlarged and improved, and Mr F. A. Sowter undertook the Editorship'; the first volume of the *Transactions of the British Bryological Society* was published under Sowter's editorship in 1947. 'A feeling was also expressed at the Meeting that some members would like to contribute observations on some aspect of the biology of the Bryophyta so as to widen the range of work within the Society' (Scott, 1947).

Many members of the generation which came to the fore in the early post-war years had biological training, and the quality of their published papers did much to raise the standards of the Society. The newly founded *Transactions* (renamed *Journal of Bryology* in 1972) provided a vital outlet for the publication and dissemination of this work. The check-lists of British mosses and liverworts were revised, Warnstorf's classification of *Sphagnum* abandoned and the nomenclature of other genera brought into line with that used in other countries (Richards & Wallace, 1950; Jones, 1958). An identification guide to the commoner mosses and liverworts, aimed particularly at beginners, was written by E. V. Watson (1955). Dixon and Macvicar's *Handbooks* became very outdated but it was many years until they were superseded. A. C. Crundwell and E. F. Warburg planned a new moss flora but Warburg died, prematurely, before it could be written. Eventually A. J. E. Smith's *Moss Flora of Britain and Ireland* (1978) provided the much-needed replacement for Dixon. His new hepatic flora, *The Liverworts of Britain and Ireland*, was published in 1990. Smith's two bryophyte floras provide the classification and nomenclature adopted in this atlas.

Species new to science, or new to the British Isles, were added at a rate which far exceeded that in the inter-war years (Watson, 1985). Some had simply been overlooked because they grew in remote places (e.g. *Dicranodontium subporodictyon*), were inconspicuous (e.g. *Leptobarbula berica*) or were closely allied to existing species (e.g. *Metzgeria temperata*). Others resulted from the taxonomic revision of troublesome groups. The recognition of the taxonomic value of rhizoid tubers led to the discovery of numerous new or neglected species of ephemeral mosses (e.g. *Dicranella staphylina*). A number of introductions were detected (e.g. *Lophocolea bispinosa*) and the spread of *Campylopus introflexus*, in particular, was closely monitored (Richards, 1963; Richards & Smith, 1975).

The procedure for registering new vice-county records was tightened up in 1946: henceforth additions were only accepted if supported by a voucher specimen. The increase in bryological activity was reflected in the number of new vice-county records, which rose during the 1950s to reach a peak in 1964. The *Census Catalogues* were revised by Warburg (1963) and J. A. Paton (1965a); in 1981 a single catalogue covering both mosses and liverworts was published for the first time (Corley & Hill, 1981).

The number of county floras also increased (Pearman, 1979). Three were published in the first volume of the *Transactions;* most subsequent floras appeared in this journal or in conjunction with vascular plant floras. The post-war format was similar to that of the pre-war years, and not radically different from that of Ray's *Catalogus* of 1660. Greater attention tended to be paid to the ecology of species: this emphasis is noticeable in the floras of Kent (Rose, 1949) and Berkshire and Oxfordshire (Jones, 1952, 1953) and in several subsequent floras, as well as in analogous papers such as Paton's Bryophyte Flora of the Sandstone Rocks of Kent and Sussex (1954). In the 1960s the division of counties into districts was largely replaced by a division based on the Ordnance Survey National Grid. The first flora based on 10-km grid squares is Paton's Bryophyte Flora of South Hants (1961). The publication date is remarkable, as it preceded by one year the publication of the *Atlas of the British Flora* (Perring & Walters, 1962), which popularized the use of the National Grid in biological recording, and by three the first vascular plant flora to cite records in grid squares, Perring *et al.*'s

Flora of Cambridgeshire. J. C. Gardiner (1981) achieved a greater degree of precision by illustrating the distribution of Surrey bryophytes in 5-km squares.

In recent years some vascular plant floras have become reduced solely to atlases which show the recent distribution of species on a 5-km or 2-km square ('tetrad') scale. Fine-scale mapping is valuable in documenting the detailed distribution and frequency of species. The effort needed to map plants on a tetrad scale is, however, so great that other features of floras have tended to be neglected. The critical evaluation of historic records, the description of the precise ecological requirements of a species and the recording of reproductive performance have been valuable features of bryophyte county floras. It would be a pity if they were hastily discarded.

THE BBS MAPPING SCHEME

Objectives

The Mapping Scheme was set up in 1960 to map the 10-km square distribution of the bryophyte flora of the British Isles. Two changes during the 30 years that the scheme has been in operation have led to the modification of this objective. Concern about the conservation of the bryophyte flora has grown and has led to an increasing demand for information on the current distribution of taxa. Secondly, developments in information technology have made it feasible to computerize the detailed data associated with each record. The final objectives of the scheme were:

(1) To collect data on the distribution of British bryophytes at a 10-km square scale;

(2) To use the data collected to create a database on the geographical distribution of British bryophytes; and

(3) To produce from this database maps of the British bryophytes at the 10-km square scale, distinguishing recent from older records.

Progress of the scheme

The Mapping Scheme was one of the first schemes modelled on the pioneer Botanical Society of the British Isles survey for the *Atlas of the British Flora* (Perring & Walters, 1962). The progress of the scheme is outlined below.

1960   Mapping scheme launched, with H. L. K. Whitehouse as Director and A. J. E. Smith as Secretary. The post of Director proved to be unnecessary, and the scheme has been run by A. J. E. S. as Mapping Secretary.

1963–78   Distribution maps were published annually in the *Transactions of the British Bryological Society* (later *Journal of Bryology*). A total of 224 maps were compiled by 29 authors, the main contributors being H. J. B. Birks, J. A. Paton, D. A. Ratcliffe and A. J. E. Smith. Most maps were of rarer species, and they were usually accompanied by a short explanatory note.

1978   Maps of 104 species were published in a *Provisional Atlas* (Smith, 1978b). These maps were mainly of widespread species, and were not accompanied by explanatory notes.

1982   A trial sample of field cards was processed by the Biological Records Centre (BRC).

1983   M. O. Hill was appointed Assistant Mapping Secretary.

1985–86   Detailed data for Cambridgeshire, Cornwall, Huntingdonshire and Surrey were processed. Sites of conservation interest were identified from the resulting database and reports produced for each county. N. G. Hodgetts and Miss A. E. Newton were employed by the Nature Conservancy Council (NCC) to work on this project under C. D. Preston at BRC.

1986–87   Further data were processed by N. G. Hodgetts and M. M. Yeo, working for NCC at BRC.

1987–89   NCC funding was obtained for two posts for two years, to complete the data processing at BRC and produce the distribution maps. R. A. Finch (1987–89), Miss C. M. Hine (1987–88) and Miss N. M. Gomes (1988–89) were employed to carry out this work. By the end of 1989, over 770,000 records had been processed.

1987   A steering committee was set up to oversee the completion of the Atlas, chaired by P. T. Harding (BRC) with M. O. Hill, R. E. Longton (Secretary, BBS), C. D. Preston and A. J. E. Smith as members. The committee met at approximately six-monthly intervals until 1989.

Maps illustrating the progress of recording have been published in the *Transactions of the British Bryological Society/Journal of Bryology* **5**, 166, 1966; **5**, 601, 1968; **6**, 331, 1971; **8**, 123, 1974; **10**, 73, 1978 and **11**, 169, 1980. A coverage map was also published in the *Provisional Atlas* (Smith, 1978b).

Of the editors of the Atlas, A. J. E. Smith has been responsible for running the Mapping Scheme for the BBS since 1960, C. D. Preston for the work of data processing and map production at BRC and for writing the introductory material, and M. O. Hill for editing the explanatory notes on individual species, preparing maps of environmental data and typing up the entire text into a word-processor for submission to the printer in machine-readable form.

SOURCES OF RECORDS

Data acquired during the mapping scheme were derived from a variety of sources which are described below.

Field survey

The survey of 10-km squares has been one of the main activities of BBS members since the inception of the Mapping Scheme, and these data provide the majority of the records available to us. In addition to the fieldwork of individual members, recording has been the primary object of the spring and summer field meetings of the Society, and many of the venues were chosen with a view to recording in under-worked areas. Reports on individual meetings were published in the *Transactions of the British Bryological Society* and the *Journal of Bryology* until 1974, and have subsequently appeared in the Society's *Bulletin*. M. A. Pearman's (1979) bibliography cites all field meeting reports for the period 1954–78.

Vice-county or county data-sets

Detailed data on the distribution of bryophytes in some vice-counties are held by individual BBS members, often because they have published or are preparing a county flora. The data-sets for Berwickshire, Cambridgeshire, Cornwall, Dorset, Hertfordshire, Huntingdonshire and North Wales were made available to the Mapping Scheme. Summary data were provided by members holding records from other counties.

Literature records
Records have been extracted from numerous literature sources, including some county floras and a selection of other publications. A particular effort has been made to extract records from areas that are badly covered by recent fieldwork.

Herbarium records
Draft liverwort maps were supplied to D. G. Long, who systematically extracted additional Scottish records for many species from the herbarium of the Royal Botanic Garden, Edinburgh (E). The compilers of the maps published in the BBS *Transactions* and *Journal* often extracted records from the major national herbaria, and records from Edinburgh, Glasgow University (GL) and the National Museum of Wales (NMW) were included in maps published in the *Provisional Atlas*. Otherwise, it has not been possible to extract records from herbaria, although some records have been received indirectly from members who have compiled data for individual counties.

Records resulting from taxonomic revision
Records of some species have been systematically revised in recent years, in some cases for the latest edition of the *Census Catalogue* (Corley & Hill, 1981). Lists of specimens examined have been made available to the Mapping Scheme. In the case of some critical taxa, only records that have received expert verification have been accepted for mapping: a comment to this effect has then been added to the notes on the species.

Records collected for published maps
Authors preparing the data for maps published in the *Transactions* or *Journal* between 1963 and 1978 usually searched herbarium and literature sources for records of the species for which they were responsible. In most cases these records were deposited with the Mapping Secretary and have been incorporated into the updated maps.

New vice-county records
For many years new vice-county records accepted by the BBS Recorders of Mosses and Hepatics were not automatically incorporated into the data being collected for the Mapping Scheme, the assumption being that they would be forwarded independently by their discoverer. From 1983 (mosses) and 1987 (liverworts) all records have been copied to the Scheme by the Society's Recorders.

'Missing' vice-county records
After the records from the above sources had been computerized, the vice-county distribution of each species, derived from the 1981 *Census Catalogue* with published additions and amendments, was checked against the data received by the Mapping Scheme. Computer-generated lists of vice-counties in which species had been

recorded, but for which there were no Mapping Scheme records, were sent to M. M. Yeo, who endeavoured to trace the missing records. In most cases these were traced and added to the database. Where a record for a vice-county cannot be localized, the species will be deleted from the county list.

RECORD CARDS

Most of the data collected during the Mapping Scheme have been extracted on to standard record cards, which are now held in the Biological Records Centre archive. Three types of card have been used.

Field cards
A card bearing the code numbers and abbreviated names of all but the rarest species was printed at the start of the scheme (Figs 1 and 2). The names of additional species could be written on the card. In subsequent reprints, the list and numbering of species remained unchanged. Field cards were used for recording in the field, for compiling summary lists or 'master cards' for individual 10-km squares, and for extracting data from literature sources.

Single species cards
Cards designed to take multiple records of a single species were used for extracting data from herbaria, literature sources or unpublished manuscripts.

Individual record cards
Cards on which a detailed individual record can be entered have been used for field, herbarium and literature records.

Data from separate field cards for a 10-km square were copied by the Mapping Secretary on to a master card for that square. It was originally envisaged that these master cards would provide the basic data for mapping. However, improvements in computer technology during the scheme made it feasible to incorporate the data on each original field card separately into the database.

In addition to data on cards, records from a number of manuscripts have been input directly. Photocopies of these manuscripts are retained in the BRC archive.

ALLOCATION OF RECORDS TO GRID SQUARES

Data collected specifically for the Mapping Scheme have been recorded in known grid squares and these grid references have been supplied by the recorder. Grid references are also available for some other recent records, made by bryologists who habitually note this information. Other modern records, and all pre-1950 records, lack grid references and have been allocated to the appropriate 10-km square on the basis of the locality information that accompanies them. Some records could not be allocated with certainty to one grid square as they were made at localities lying on the border of two squares or at the junction of four. They were then assigned to a grid square according to the following rules.

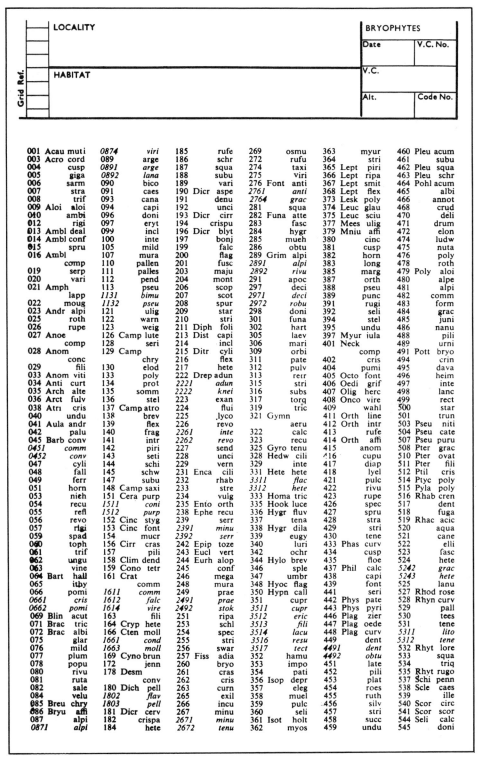

Fig. 1. The field card used during the BBS Mapping Scheme (side 1). The card was reprinted several times during the course of the scheme, with minor changes in design but the same species list.

Biological Records Centre   October 1972

Code	Genus	Species
546		pauc
547		pusi
548		recu
549		trif
551	Sema	
		nova
554	Spha	
		comp
555		cont
556		cusp
557		fimb
558		fusc
559		girg
560		imbr
562		mage
563		moll
564		nemo
566		palu
567		papi
568		plum
569		pulc
570		quin
571		recu
573		rube
574		russ
575		squa
576		stri
578		subs
5781		*auri*
5783		*inun*
5785		*subs*
579		tene
580		teres
581		warn
582	Spla	
		ampu
583		ovat
588	Tetr	brow
589		pell
591	Tetr	mnio
592	Tham	
		alop
594	Thui	abie
595		deli
596		hyst
597		phil
599		tama
602	Tort	flav
604		incl
605		infl
606		niti
607		tort
610	Tort	inte
611		laev
612		lati
613		marg
614		mura
6142		*mura*
6143		*rupe*
616		papi
617		prin
618		ruralif
619		ruralis
620		subu
624	Tric	brac
6241		*brac*
624		*coph*
6243		*litt*
625		cris
627		sinu
628		tenu
630	Ulot	amer
631		bruc
632		crispa
634		drum
635		ludw
637		phyl
639	Weis	cont
640		crispa
641		crispat
642		micr
645		ruti
650	Zygo	cono
653		viri
6531		*stir*
6532		*viri*
6533		*vulg*
681	Adel	deci
683	Anas	orca
684	Anas	doni
686	Anth	jula
688	Anth	husn
689		laev
690		punc
691	Apha	
		micr
692	Barb	atla
693		atte
694		barb
695		floe
696		hatc
698		lyco
700	Bazz	pear
701		tric
702		tril
703	Blas	pusi
704	Blep	tric
705	Caly	argu
706		fiss
707		muel
708		nees
7081		*meyl*
7082		*nees*
709		spha
711		tric
714	Ceph	bicu
7141		*bicu*
7142		*lamm*
715		cate
716		conn
718		leuc
719		loit
720		macr
721		medi
727	Ceph	
		hamp
730		pear
731		rube
732		star
736	Chan	seti
737	Chil	pall
738		poly
7381		*poly*
7382		*rivu*
739	Clad	flui
740		fran
741	Colo	calc
742		minu
743		rose
744	Colu	caly
745	Cono	coni
747	Dipl	albi
748		obtu
749		taxi
750	Doui	ovat
751	Drep	
		hama
753	Erem	myri
754	Foss	angu
755		caes
758		fove
762		pusi
763		wond
764	Frul	dila
765		frag
766		germ
767		micr
768		tama
772	Gymn	infl
774	Gymn	
		conc
776		cren
777		obtu
779	Harp	ovat
780	Harp	flot
781		scut
782	Herb	adun
783		hutc
784	Hygr	laxi
785	Jame	autu
786		carr
788	Jubu	hutc
790	Leio	bade
791		bant
794		muel
796		turb
797	Leje	cavi
801		lama
8011		*azor*
8012		*lama*
803		pate
804		ulic
805	Lepi	pear
806		pinn
807		rept
808		seta
809		sylv
810		tric
811	Loph	bide
812		cusp
813		frag
814		hete
815	Loph	alpe
816		bicr
818		exci
819		inci
820		long
823		porp
824		vent
826	Lunu	cruc
827	Marc	poly
8271		*alpe*
8272		*aqua*
8273		*poly*
828	Marc	
		mack
829	Mars	adus
830		alpi
831		aqua
834		emar
835		func
837		spha
839		stab
840		ustu
841		vari
842	Mast	
		wood
843	Metz	conj
844		frut
845		furc
846		hama
847		pube
849	Moer	blyt
850		flot
852	Myli	
		anom
853		cune
854		tayl
856	Nard	
		comp
857		geos
858		scal
859	Nowe	
		curv
860	Odon	
		denu
863		spha
864	Pall	lyel
865	Pedi	inte
867	Pell	epip
868		fabb
869		nees
870	Peta	ralf
872	Plag	aspl
8721		*aspl*
8722		*majo*
873		punc
874		spin
875		trid
876	Plec	hyal
877		obov
878		paro
880	Pleu	albe
881	Pleu	purp
882	Pore	cord
883		laev
884		pinn
885		plat
886		thuj
887	Prei	quad
888	Ptil	cili
889		pulc
890	Radu	aqui
892		comp
894		lind
896	Rebo	hemi
897	Ricc	incu
898		lati
899		mult
900		palm
901		ping
902		sinu
903	Ricc	beyr
908		crys
909		flui
910		glau
915		soro
916		warn
917	Ricc	nata
918	Sacc	viti
919	Scap	aequ
920		aspe
922		comp
924		curt
927		grac
929		irri
930		nemo
931		nimb
932		orni
936		suba
937		ulig
938		umbr
939		undu
940	Sole	atro
9402		*spha*
942		cord
943		cren
944		pumi
946		spha
947		tris
952	Sphe	minu
954	Targ	hypo
956	Tric	tome
957	Trit	
		exsecta
958		exsectif
960		quin

Other Species

Biological Records Centre   October 1972

RP8

Fig. 2. The field card used during the BBS Mapping Scheme (side 2).

(1)   If the species had been definitely recorded from one of the possible squares, the record was allocated to that square.

(2)   If the species was not recorded from any of the possible squares, the record was allocated to the square that seemed the most likely, on the basis of the locality data or, occasionally, the known habitat preferences of the species.

(3)   Failing all else, the record was allocated at random to one of the possible squares.

Records which were so vague that they could have come from more than two or (in the case of localities near the intersection of 10-km grid lines) more than four squares have usually not been assigned to a grid square and do not appear on the maps.

In the early years of the Mapping Scheme, Irish records were not collected in the 10-km squares of the current Irish National Grid, but on the extension of the British grid devised by D. A. Webb (1955) for Ireland. This extension grid was used by the Botanical Society of the British Isles when recording for the *Atlas of the British Flora* (Perring & Walters, 1962), because maps with the Irish grid were not then available. Records collected on the extension grid have been re-allocated to the Irish grid. If the records were localized, the locality information was used to place them in the correct Irish grid square. Unlocalized records have been allocated to the most appropriate square, using a standard conversion table. Although some unlocalized records will inevitably be misplaced, all must lie in the correct square or in a square adjacent to it.

A number of conventions followed by the *Atlas of the British Flora* were initially adopted by the BBS. Records from all four grid squares in the Isles of Scilly were, for administrative convenience, placed in a single square (initially SV90, later SV91). In this Atlas they are plotted in the correct grid square. The *Atlas of the British Flora* also employed a conventional grid reference for each of the Channel Islands, as gridded Channel Island maps were not available. We have plotted localized Channel Island records in the correct squares of the UTM grid, which now appears on maps of the islands, but some of our records are attributed only to an island and these have been plotted in conventional grid squares (WA50 for Alderney, WV27 for Guernsey, WV38 for Herm and WV65 for Jersey; the whole of Sark lies in WV47).

Recorders may also have been influenced by the *Atlas of the British Flora* convention which allocated records from coastal squares with a small area of land to adjacent squares with a greater land area. Wherever possible we have plotted records in the correct grid square, irrespective of its land area, but we may have been unaware of such transfers or have lacked the data to correct them.

ALLOCATION OF RECORDS TO VICE-COUNTIES

Some field recorders included records from more than one vice-county on a record card for a 10-km square. These records have been allocated to the vice-county which occupies the largest area of the square.

DATING OF RECORDS

The original intention of the Mapping Scheme was to publish maps with all records plotted as a single symbol, regardless of date. Such maps were produced for the *Provisional Atlas* (Smith, 1978b). In the later years of the Scheme it was decided to distinguish recent records, as opinion in ITE and NCC strongly favoured this. Records made in or after 1950 are therefore distinguished on the maps from those made before. The choice of cut-off date is necessarily a compromise between the desire to obtain as accurate a picture as possible of the current bryophyte flora and the time needed to survey the grid squares adequately. A date later than 1950 would present a misleading picture as the recent coverage would be inadequate.

The decision, made at a late stage, to distinguish two date classes has led to three problems. A few county data-sets were started before 1950 and it is not now possible to distinguish the small minority of records made before 1950 from the majority recorded subsequently. Records from North Somerset dating from 1949 onwards (Appleyard, 1970) were assumed to refer to the period from 1950. Records mapped by A. G. Side (1970) have also been treated as recent, although they include records derived from F. Rose's fieldwork in the county from 1943 onwards. Secondly, undated records have been received. If the records were attributed to a recorder, it has usually been possible to date them approximately from a knowledge of the period when that recorder was active. However, undated data provided from 10-km squares in summary form give no details of recorder and have perforce to be left undated and mapped with the pre–1950 records. The third problem is more difficult to assess. Recorders thinking that an Atlas would not distinguish date classes may not have been too scrupulous in separating modern records from old records. Some old records may therefore have been added to field cards with recent dates.

DATA PROCESSING AND VALIDATION

The following steps were taken to process and check the data.

(1) Localities were checked manually against the relevant Ordnance Survey map to ensure that the grid reference given was correct.

(2) Records were then entered into the computer. The following data, if available, were placed on file:

> Species code number
> Grid reference
> Vice-county
> Locality
> Recorder
> Determiner (if different from recorder)
> Date
> Source of record: field record, herbarium record (with herbarium code) or literature record (with code for source).

(3) A print-out of records was checked against the original data and input mistakes were corrected.

> The records were then passed through a suite of computer programs to validate them (steps 4–6).

(4) Species code numbers were checked against a standard list. Code numbers of taxa which are no longer recognized were corrected at this stage (e.g. records of *Marchantia polymorpha* var. *aquatica* were renumbered as *M. polymorpha* sensu stricto).

(5) The 10-km square was checked for each record against the vice-county to ensure that they were compatible. Records showing a discrepancy were examined and corrected by comparison with the original record card.

(6) The records of each species were checked against its vice-comital distribution (based on Corley & Hill (1981), with published additions and amendments). Records from vice-counties for which a species was not officially recorded were deleted. In this way, dubious and impossible records, perhaps resulting from mistakes in crossing abbreviated names on field cards (cf. Allen, 1981), were removed. A minority of deleted records were doubtless correct records, made by recorders who did not appreciate that they were new to the vice-county and hence did not submit a voucher specimen to the relevant recorder. Vice-counties from which a species had been recorded, but for which no records were available, were also identified and the records added (see 'Sources of records' above).

At this stage draft maps were printed and records listed. These listings and maps were scrutinized and critically revised. Additional records, including those from 'missing' vice-counties, were added and checked. The final maps were then plotted.

AVAILABILITY OF RECORDS

The records will be held by BRC on the ORACLE database management system and, on completion of the three volumes of the *Atlas of the Bryophytes of Britain and Ireland*, will be available for use by research workers. The BBS, ITE and NCC wish to encourage the use of this database in taxonomic, biogeographical or ecological research, in assessing the conservation needs of bryophytes and in monitoring change in the bryophyte flora. The archive of record cards and associated manuscripts will also be maintained at the Biological Records Centre and may be consulted by visitors with a *bona fide* interest. All enquiries should be addressed to the Biological Records Centre, Monks Wood Experimental Station, Abbots Ripton, Huntingdon, PE17 2LS, Great Britain.

FUTURE RECORDING

The BBS plans to continue to record bryophytes at a 10-km square scale, in order to build on the foundation laid by the Atlas. Experience with other taxonomic groups shows that the publication of an Atlas for any taxonomic group is the most effective way of identifying recording errors and drawing attention to under-recorded areas. It is desirable that errors should be corrected and under-recorded areas surveyed.

Records in a database are much more useful if they are localized, attributed to a particular recorder and accurately dated than if they are simply lists of records for 10-km squares in a broad date-class. It is particularly important that future records should be dated if changes in the bryophyte flora, such as those which may result from land-use change or climatic change, are to be identified and monitored. Enquiries about recording should be directed to the Recording Secretary, British Bryological Society, c/o Department of Botany, National Museum of Wales, Cardiff, CF1 3NP, Great Britain.

EVENNESS OF THE SURVEY

Variation in recording of species

Bryophytes vary in the ease with which they can be recognized and recorded. The following groups tend to be under-recorded.

(1) Species which are small and inconspicuous, or which tend to occur as scattered stems rather than as tufts or mats (e.g. *Diplophyllum obtusifolium, Lejeunea hibernica*). The subterranean *Cryptothallus mirabilis* is particularly likely to be overlooked.

(2) Local species which resemble more widespread taxa (e.g. *Diplophyllum taxifolium, Jungermannia subelliptica*).

(3) Species which can be identified only if gametangia or sporophytes are present (e.g. *Jungermannia hyalina, Pellia neesiana*).

(4) Species in taxonomically critical species' pairs or species' groups (e.g. *Cephaloziella* spp., *Scapania curta, S. lingulata, S. scandica*).

(5) Species which can be identified only by specialist techniques (*Pellia borealis*).

(6) Species which have only recently been recognized in the British Isles (e.g. *Fossombronia fimbriata*) or which, although known, have until recently been confused with related plants (e.g. *Metzgeria fruticulosa*).

There is considerable overlap between these groups: species recently added to the British flora, for example, tend to be small and inconspicuous or to belong to critical groups.

The notes accompanying the individual maps indicate which species are thought to be under-recorded.

Variation in geographical coverage

We hope to include a detailed analysis of the geographical coverage in the final volume of the Atlas. In the meantime, the following notes highlight areas which are particularly well or particularly badly recorded.

In England, the south-eastern counties, where most bryologists live, tend to be well recorded. Exceptions, where coverage is less good, include Norfolk and East Suffolk. In south-west England, Cornwall and Somerset are exceptionally well covered. Coverage of some of the bryologically unpromising counties of the English Midlands, including Shropshire, Staffordshire and Nottinghamshire, is poor. Some areas of northern England, including the eastern side of Yorkshire and the land bordering the Solway Firth, are also poorly recorded.

There are some English counties where there has been little recent fieldwork, but which were quite well covered before 1950. These include Northamptonshire (where old records are available for mosses but not for liverworts), Leicestershire and South Lancashire.

North Wales is very well recorded; South Wales has been less intensively worked.

South-west Scotland (Dumfries, Galloway and Ayr) is poorly recorded. Kintyre is also poorly recorded, but much of western Scotland, including Argyll and Westerness, is well covered, and Skye is exceptionally well worked. The Lothians and the bryologically rich Breadalbane range in Perthshire are well worked. Many parts of the Highlands without high mountains are poorly known, including areas of Inverness and Ross. Coverage of eastern Scotland is patchy: east Fife is poorly recorded, Angus well recorded. The area of north-east Scotland which includes Kincardineshire, Aberdeenshire and Banffshire is undoubtedly the worst-recorded area on the mainland of Great Britain, and appears blank on most maps. Caithness and East Sutherland are poorly known. Shetland is well recorded but Orkney is not. In the Outer Hebrides, Lewis is poorly recorded.

In Ireland recording has been handicapped by the extreme shortage of resident bryologists. Some recording has been undertaken by visitors, but they naturally tend to gravitate towards the species-rich western seaboard. Consequently Kerry, Galway and Sligo are well recorded. Elsewhere well worked areas include Fermanagh, Meath, Tyrone and Wicklow.

MAPS OF LIVERWORT
DISTRIBUTIONS

EXPLANATION OF MAPS AND ACCOMPANYING NOTES

Records are mapped in the 10 × 10-km squares of the Ordnance Survey National Grid in Great Britain and in the Ordnance Survey/Suirbheireacht Ordonais National Grid in Ireland. Records from the Channel Islands are mapped in the 10 × 10-km squares of the Universal Transverse Mercator Grid.

The symbols used are

○   Record made before 1950, or undated
●   Record made in or after 1950.

The numbering and nomenclature of species is that of *The Liverworts of Britain and Ireland* (Smith, 1990). If the name used by Smith (*op. cit.*) differs from that used by Corley & Hill (1981), the latter is cited as a synonym and included in the index. A fuller synonymy is given in Smith's *Liverworts*.

Each map is accompanied by notes on the taxon mapped. The first paragraph describes the habitat. The altitudinal range is given (in metres). 'Lowland' denotes altitudes below 300m. The paragraph ends with a formula indicating the number of 10-km grid squares in which the taxon is mapped. GB 18+11*, IR 6+8* indicates that the plant has been recorded in or after 1950 in 18 grid squares in Great Britain and 6 in Ireland. There are pre-1950 or undated records from an additional 11 squares in Great Britain and 8 in Ireland. Channel Island records are not included in these totals.

The second paragraph describes the sexuality of the taxon, the frequency of sporophytes and indicates whether it has specialized means of vegetative spread, e.g. by gemmae (usually few-celled, more or less undifferentiated structures), bulbils, tubers, branchlets or fragile leaves. The following terms may be used to describe sexuality:

Sterile: antheridia (male sex organs) and archegonia (female sex organs) not produced;

Dioecious: antheridia and archegonia borne on separate plants; and

Monoecious: antheridia and archegonia borne on the same plant.

Some authors have distinguished separate categories of monoeciousness:

Autoecious: antheridia and archegonia borne in separate inflorescences;

Paroecious: antheridia naked in the axils of the leaves immediately below the female inflorescence; and

Synoecious: antheridia and archegonia borne in the same inflorescence.

The third paragraph describes the distribution of the taxon outside the British Isles. Information on the extra-European distribution of British species is sometimes of doubtful accuracy. Species reported from other continents may not be the same as the European plant given that name, or the European plant may be known by another name elsewhere. Readers should bear in mind this element of uncertainty.

Additional comments may be given in a fourth paragraph.

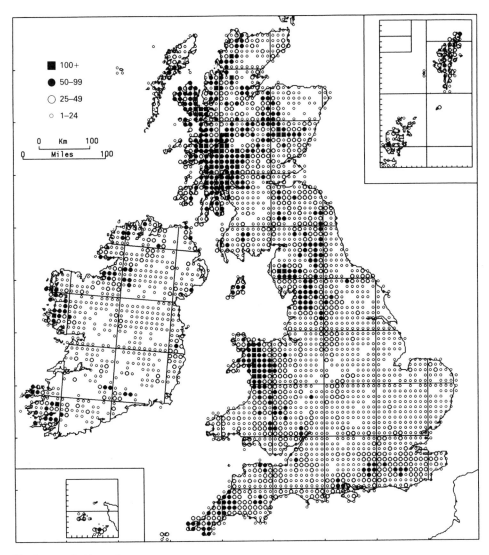

Totals recorded in 10-km squares

The map shows the number of liverwort species recorded per 10-km square. Records have been counted without regard to date, including old records as well as new ones. Infraspecific taxa have not been counted separately.

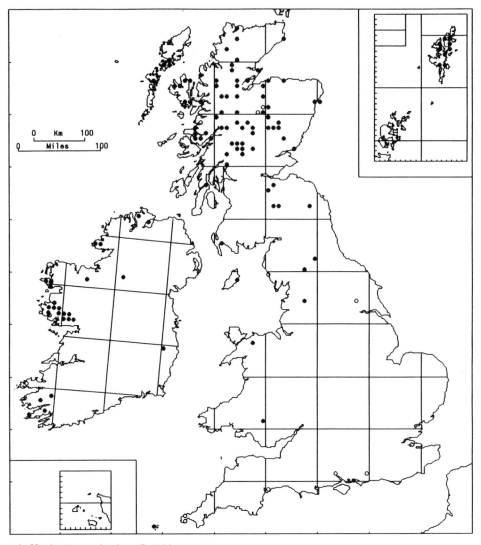

1/1. **Haplomitrium hookeri** (Sm.) Nees

The only British member of the Calobryales; it occurs at low altitudes in small quantities on slightly basic sandy or gravelly substrata in a range of habitats including dune-slacks, lake, stream and river margins, tracks and flushes on heaths, the floors of old quarries, and on detritus in the vicinity of dams. Frequent associates include *Blasia pusilla*, *Fossombronia fimbriata*, *F. incurva*, *Pellia epiphylla*, *Riccardia incurvata*, *Scapania* spp., *Anomobryum filiforme* and *Pohlia* spp. At higher altitudes in the Scottish Highlands restricted to bryophyte-rich turfy ledges and slopes, mainly with a north or east aspect, in corries. 0–1070 m (Cairn Gorm). GB 90+11*, IR 25+2*.

Dioecious; male and female plants and sporophytes frequent in the summer months. In northern Britain the aerial parts often die back during the winter and an extensive system of underground axes acts as the organ of perennation (Grubb, 1970). Gemmae absent.

A northern species widely distributed in Europe from the Alps and Tatra Mts north to Spitsbergen and east to European Russia. Scattered localities in Greenland and N. America from New England to Alaska.

It appears to be extremely local throughout its range and may well be most abundant in Britain.

J. G. Duckett

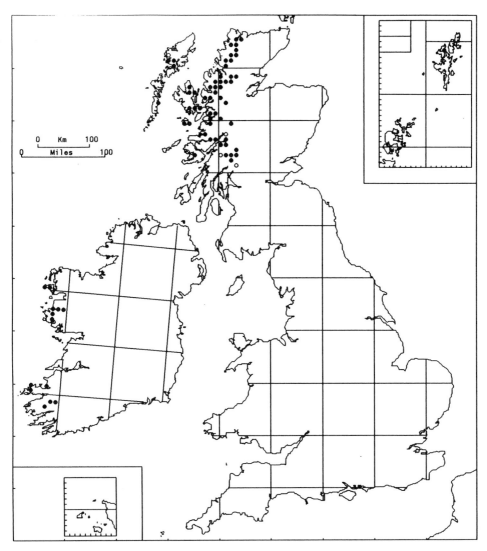

2/1. Mastigophora woodsii (Hook.) Nees

This hyperoceanic, montane species is locally frequent in parts of the N.W. Highlands of Scotland. It is a member of a distinctive mixed hepatic assemblage with other large upland species, including *Bazzania tricrenata*, *Herbertus aduncus*, *Plagiochila carringtonii*, *Pleurozia purpurea* and *Scapania nimbosa*, growing amongst heather and other dwarf shrubs on well-drained, rocky slopes. It also occurs, in acidic or occasionally mildly basic conditions, in upland grassland, block-screes, cliff-faces and rock ledges. Most localities have a northerly to easterly aspect and all have a cool and extremely humid climate. Mainly at altitudes 300–1000 m, but descending below 100 m. GB 56+4*, IR 13.

Sterile and lacking gemmae.

M. woodsii has a markedly disjunct distribution; outside the British Isles it has been recorded in the Faeroes, north-west N. America and locally in C. and E. Asia. Abundant in the eastern Himalaya, where it is largely a woodland species.

T. H. BLACKSTOCK

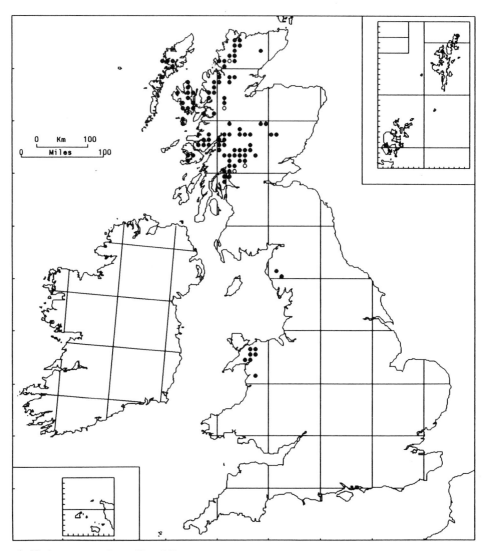

3/1. **Herbertus stramineus** (Dum.) Trev.

This species is mainly confined to montane base-rich habitats where it can be locally frequent. It commonly occurs as pure tufts or mixed with other bryophytes on damp, shady basic cliff-ledges, generally N.- or E.-facing and often where water drips or spray falls at times. It is often associated with 'tall herbs' such as *Sedum rosea* and *Saussurea alpina* and with bryophytes such as *Mastigophora woodsii*, *Ctenidium molluscum* and *Leptodontium recurvifolium*. It also occurs more rarely in damp, rather open species-rich grass or herb communities on steep, moist but well-drained N.- or E.-facing slopes. In some localities it is accompanied by *H. aduncus* on basic rocks but it does not occur with the other North Atlantic large leafy hepatics in oligotrophic habitats. All localities have a cool, moist climate, and it evidently tolerates moderately prolonged snow cover. Extends from 20 m (North Harris) to 1180 m (Ben Lawers). GB 102+4*.

Dioecious; both sexes occur in Britain but sporophytes have not been found.

A North Atlantic species, known from Faeroe Islands, W. Norway, Iceland and the unglaciated part of N. Alaska.

H. J. B. Birks & D. A. Ratcliffe

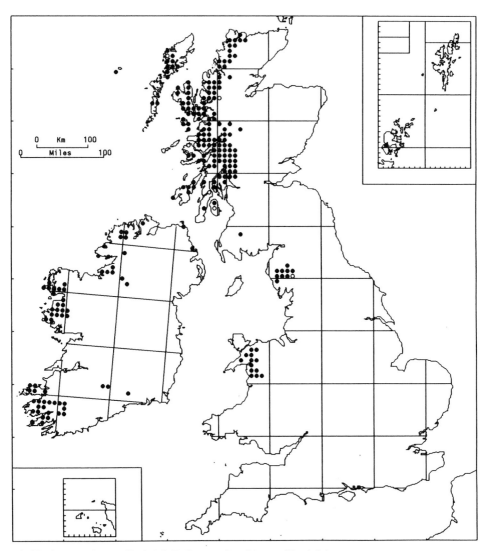

3/2. **Herbertus aduncus** (Dicks.) S. F. Gray ssp. **hutchinsiae** (Gott.) Schust.

Most frequent in oligotrophic dwarf-shrub communities on steep, well-drained, block-strewn slopes in N.- or E.-facing corries and on shaded acid or mildly basic cliff-ledges, where it grows in dense pure cushions or mixed with other large liverworts, including *Anastrepta orcadensis*, *Bazzania tricrenata*, *Plagiochila spinulosa*, *Pleurozia purpurea*, *Scapania gracilis*, and, more rarely, *Mastigophora woodsii* and *Plagiochila carringtonii*. It also occurs in more open habitats such as stable N.-facing block-screes, and ledges by waterfalls. In W. Scotland, it descends to near sea-level on ledges in wooded ravines, on boulders and cliffs in rocky birch woods, amongst dwarf shrubs on shaded humus banks, and on steep N.- or E.-facing slopes. It does not grow in areas of late snow-lie. 0–750 m, reaching 1040 m in S.W. Ireland. GB 159+7*, IR 64+1*.

Female only; perianths rare.

Outside the British Isles, occurs very rarely in W. Norway. Ssp. *aduncus* occurs in Alaska and British Columbia, and ssp. *tenuis* in the Appalachians of eastern N. America.

Although widely distributed in the British Isles and locally abundant, its distribution at a local scale is curiously patchy; absent from several seemingly suitable hills. Appears sensitive to fire and grazing.

H. J. B. Birks & D. A. Ratcliffe

39

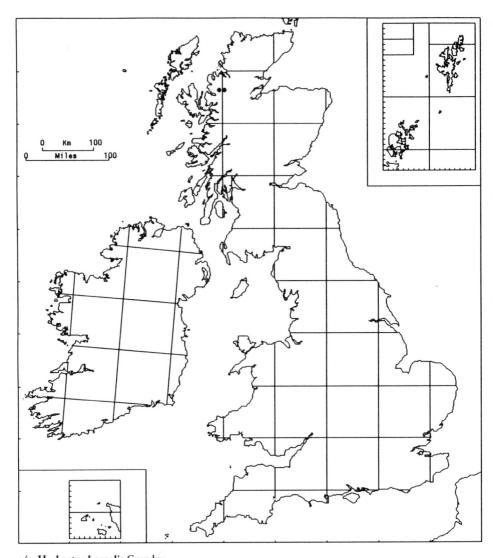

3/3. **Herbertus borealis** Crundw.

Although not described until 1970, this distinctive and robust plant has been collected from its single Scottish locality since 1868. It is locally abundant there, growing amidst open wind-exposed prostrate juniper heath on a gentle N.E.-facing slope with *Arctostaphylos uva-ursi*, *Calluna vulgaris*, *Pleurozia purpurea* and *Racomitrium lanuginosum*, and in tallish *Calluna* heath on N.- or N.E.-facing slopes. Soils are oligotrophic, humus-rich rankers developed over quartzite moraine and scree. It locally plays the role that *H. aduncus* has elsewhere in Scotland. Altitudinal range 380–550 m. It occurs in an area of cool, humid climate with at least 220 wet days a year. GB 2.

Dioecious; female inflorescence unknown, male inflorescence unknown in Scotland but present in small quantity in a Norwegian specimen.

Outside Scotland, occurs in three localities near Stavanger, S.W. Norway. These localities are much more sheltered than the Scottish station. The best-known locality is a low-lying waterfall where it grows with *Hymenophyllum wilsonii*, *Pleurozia purpurea* and *Scapania ornithopodioides*. In none of its known Norwegian localities does it occur as abundantly as it does in N.W. Scotland.

H. J. B. Birks

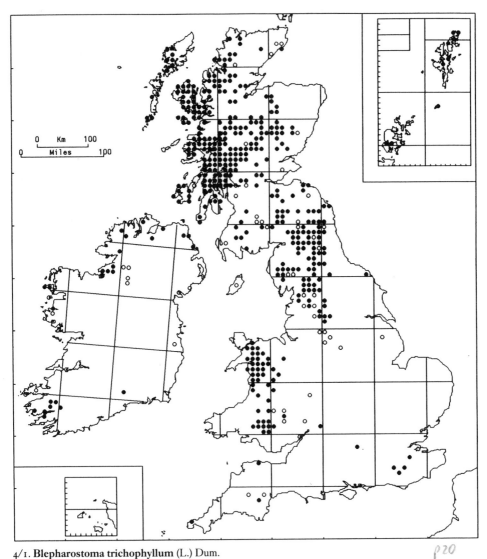

4/1. Blepharostoma trichophyllum (L.) Dum.   *p20*

Seldom present in pure tufts, *B. trichophyllum* is usually found creeping amongst other bryophytes in a variety of communities, on acidic and base-rich substrata. It is most frequent on sheltered rock-walls and ledges amongst crags and in ravines, with associates such as *Amphidium mougeotii*, *Anoectangium aestivum* and *Gymnostomum aeruginosum*, but it also occurs in flushes, by streams and on peaty moorland banks. In deciduous woodland, it grows on tree-bases, logs and within cushions of large ground-layer mosses, notably *Leucobryum* spp. 0–1205 m (Ben Lawers). GB 388+46*, IR 25+12*.

Monoecious; sporophytes occasional. Plants producing gemmae occur in Britain but are rarely recorded.

Boreal-montane in Europe, Asia and N. America, extending southwards to mountain ranges in equatorial regions.

T. H. BLACKSTOCK

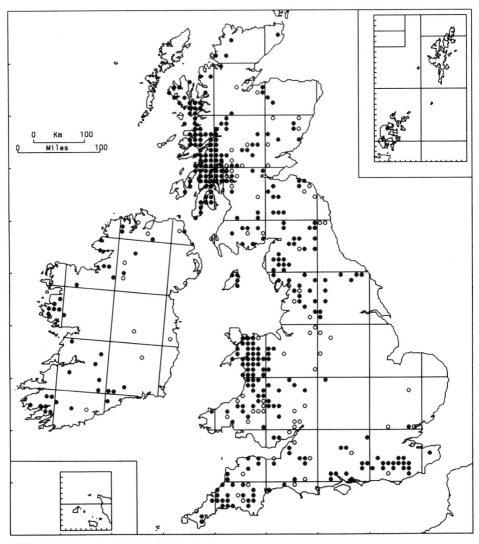

5/1. **Trichocolea tomentella** (Ehrh.) Dum.

This species is most widely distributed in damp woodlands and shaded ravines, and less frequently in wooded swamps. In favourable conditions, it forms extensive patches amongst and over other bryophytes, such as *Hookeria lucens*, *Plagiochila asplenioides* and *Rhizomnium punctatum* in seepage zones, oozes and wet stream-banks, and *Hylocomium brevirostre* and *Saccogyna viticulosa* on rock-faces by waterfalls and in other humid situations. It also occurs in more open mesotrophic flushes, mostly on hillsides in the north and west, where associates may include *Calliergon cuspidatum*, *Campylium stellatum* and *Sphagnum warnstorfii*. Mainly lowland but recorded at 700 m (Nag's Head). GB 325+71*, IR 38+10*.

Sporophytes rare and gemmae lacking.

Suboceanic in Europe; very scarce in the extreme north and the Mediterranean region. N. Africa, Himalaya, E. Asia, eastern N. America.

T. H. Blackstock

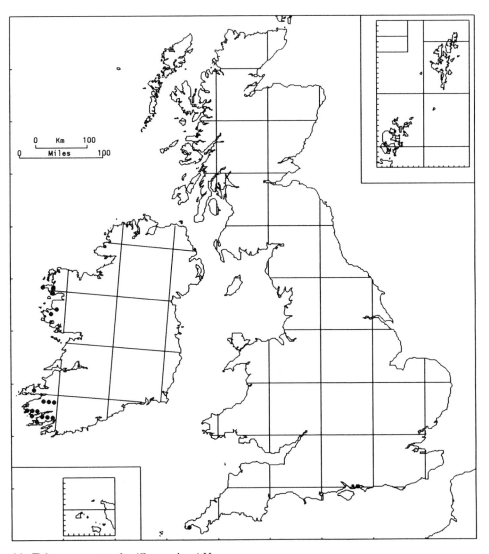

6/1. **Telaranea nematodes** (Gott. ex Aust.) Howe

This species favours moist peaty banks and peat on rocks, usually in deeply shaded, humid sites under rhododendron in woodlands and thickets on or near the coast. Very rarely, it occurs amongst *Molinia* tussocks in marshes, or in turf on coastal slopes. Lowland. GB 1, IR 17.

Monoecious, often fertile; mature sporophytes have been noted in June. Axillary bulbils are frequently produced (Paton, 1987) and evidently provide a specialized means of vegetative propagation.

A subtropical and tropical species, occurring very locally on the Atlantic fringe of W. Europe north to Ireland. Elsewhere, in Macaronesia, tropical and southern Africa, and widespread in the Americas from New York south to Peru.

T. H. BLACKSTOCK

43

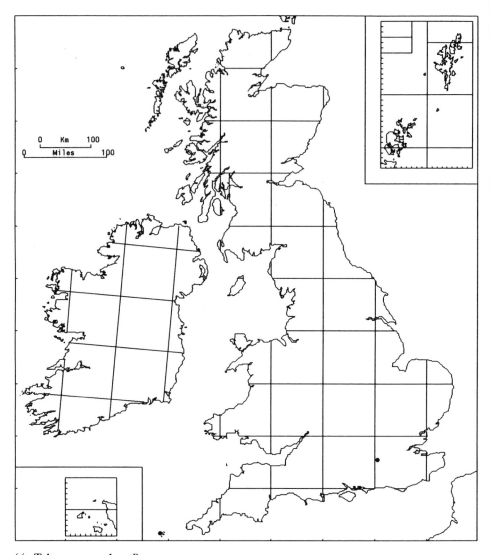

6/2. **Telaranea murphyae** Paton

On shaded peaty banks, rotten stumps and logs in a conifer wood in the Isles of Scilly, growing in pure mats or mixed with *Lepidozia reptans* and other calcifuge woodland bryophytes, including *Lophocolea semiteres*. The Surrey locality is in Wisley Gardens, where it grows on moist peat and damp sandy soil shaded by rhododendrons and ferns, also accompanied by calcifuge woodland bryophytes. GB 2.

Dioecious; male in the Scillies, female in Surrey. Gemmae absent.

Not known outside Britain.

An alien of unknown, probably austral origin, first found 1961. Both known localities are in or near botanic gardens to which numerous exotic vascular plants have been introduced.

M. O. HILL

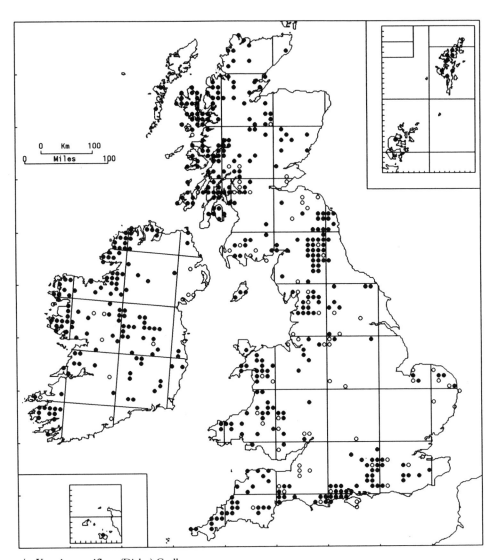

7/1. **Kurzia pauciflora** (Dicks.) Grolle

Primarily a species of saturated peat-bogs, where it usually grows amongst sphagnum or on peat, and is often associated with other similarly restricted hepatics such as *Calypogeia sphagnicola*, *Cephalozia macrostachya*, *Mylia anomala* and *Odontoschisma sphagni*. It is recorded from other mire types, including acidic flushes and amongst *Sphagnum submitens* on tussocks in fens, and also wet heaths and moors. Predominantly lowland, to 580 m (Skye). GB 353+75*, IR 133+13*.

Dioecious; sporophytes occasional. No specialized method of vegetative dispersal but, after peat disturbance, can regenerate from underground axes which exist down to below 20 cm in peat (Duckett & Clymo, 1988).

Widely distributed in N. and C. Europe, eastwards to Russia. Macaronesia, Alaska, eastern N. America.

Determinations of the three European species of *Kurzia* are often not straightforward and recorders have sometimes referred finds indiscriminately to *K. pauciflora*, the commonest but probably most ecologically distinct taxon. Although records based on erroneous and casual identification may thus be included, the maps probably provide a reasonable reflection of the distribution patterns and relative frequencies of the three species.

T. H. BLACKSTOCK

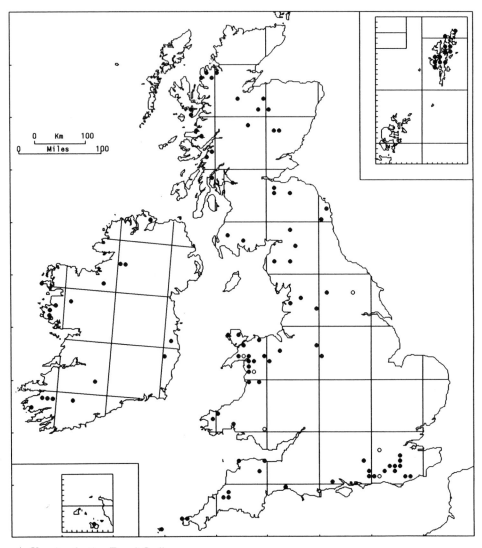

7/2. **Kurzia sylvatica** (Evans) Grolle

This *Kurzia* is most frequently recorded on steep peat-banks and moist organic soil in moorland and damp heaths, less often in bogs and occasionally in woodland. It also occurs on damp, sheltered rock outcrops, especially sandstone exposures, and moist sandy banks. The ecological range of *K. sylvatica* appears to be broadly similar to that of *K. trichoclados* in the British Isles, and coexisting populations of the two species have been reported from a number of localities. 0–600 m (Berwyn Mts). GB 90+6*, IR 17.

Dioecious; sporophytes very rare. Specialized methods of vegetative dispersal lacking.

Suboceanic in Europe; reported from Spain, France, Belgium, Holland, Germany, Austria, Denmark, Poland, Czechoslovakia. Eastern N. America.

T. H. Blackstock

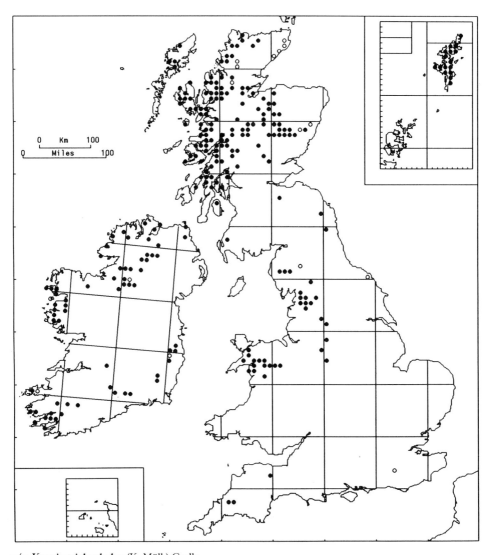

7/3. **Kurzia trichoclados** (K. Müll.) Grolle

Similar ecologically to *K. sylvatica* but generally more abundant on moorland and damp heaths in the north and west. In favourable conditions, it forms dense swollen cushions on moist peat-banks and damp peaty soil amongst rocks and on sheltered slopes. It can grow under the cover of dwarf shrubs and is also found in other shaded habitats, including woodland banks and damp ledges on steep rock-faces in ravines. 0–1000 m (Easterness). GB 183+12*, IR 64+4*.

Dioecious; sporophytes occasional. Swollen bulbils sometimes develop on senescent stems and may provide a means of vegetative propagation.

Elsewhere in Europe, reported from Scandinavia, the central montane region, Portugal, Spain. Not known outside Europe.

T. H. BLACKSTOCK

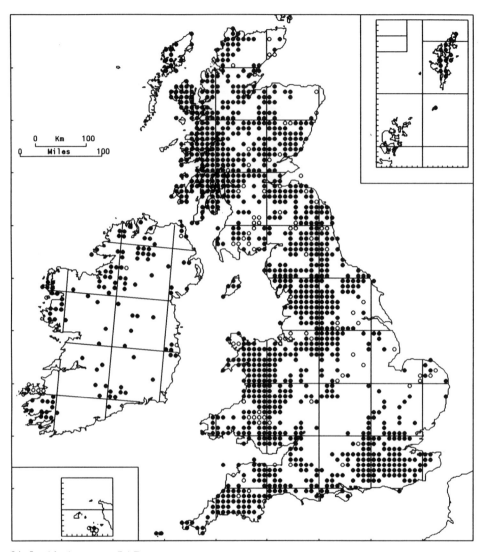

8/1. **Lepidozia reptans** (L.) Dum.

One of the most widespread and frequent hepatics found in deciduous and coniferous woodland in the British Isles. It occurs in patches and scattered amongst mosses on stumps, rotting logs, tree-trunks and bases, humus-banks and other organic substrata; less commonly it colonizes sandstone and other rocks types but rarely, if ever, base-rich exposures. It extends outside woodland into heath and moorland, especially in the west and north, where it is also found on block-strewn slopes and other sheltered rocky ground. Chiefly below 500 m, but to 1000 m on Ben Vorlich, Perthshire. GB 1145+88*, IR 116+9*.

Monoecious; sporophytes occasional to frequent.

Circumboreal. Throughout much of Europe but scarce in the extreme north and mostly absent from the Mediterranean region.

T. H. BLACKSTOCK

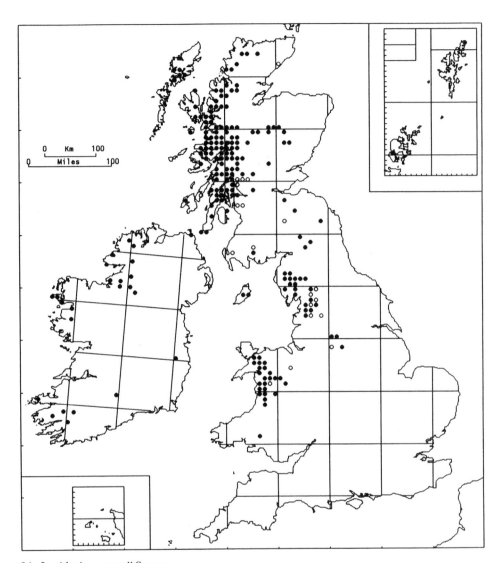

8/2. Lepidozia pearsonii Spruce

This species is found in the submontane zone, mainly in shaded localities with a northerly to easterly aspect, in dwarf-shrub heath vegetation on rocky slopes, within boulder-screes and on steep banks and ledges at the base of cliffs. Descends to low altitudes where it also occurs in rocky oak and birch woodland, wooded ravines and occasionally in Scottish pine-forest. It is often present as scattered stems amongst larger bryophytes, such as *Pleurozium schreberi* and *Sphagnum quinquefarium*, and also forming small patches sometimes with other western hepatics, including *Anastrepta orcadensis* and *Bazzania tricrenata*. 0–800 m (Glen Feshie). GB 179+22*, IR 25.

Dioecious, mostly male; female plants rare and sporophytes unknown.

Outside the British Isles, known only in Norway, mostly in the south-west coastal region.

T. H. BLACKSTOCK

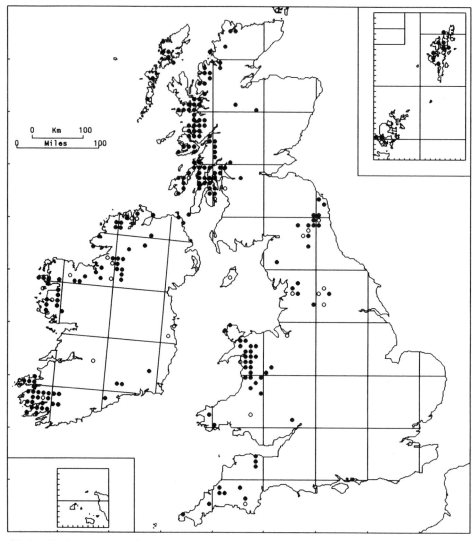

8/3. **Lepidozia cupressina** (Sw.) Lindenb.

A characteristic though rather local species of rocky oak woodland, particularly in the west, where it can form large, prominent cushions on the tops of shaded boulders and outcrops; less often found on tree-bases, stumps, rotting logs and sometimes soil. *Bazzania trilobata* is one of its common associates. It also occurs outside woodland in gullies, block-screes and other sheltered rocky ground. Mostly lowland, 0–650 m (Co. Mayo). GB 153+18*, IR 71+14*.

Male and female plants occur in the British Isles.

Very local in W. Europe outside Britain and Ireland; recorded in Norway, Germany, France, Luxemburg, Spain. Also in the Azores, Canary Islands, Madeira. Elsewhere, the *L. cupressina* complex is represented in the Caribbean region, S. America, Africa, and E. and C. Asia.

T. H. BLACKSTOCK

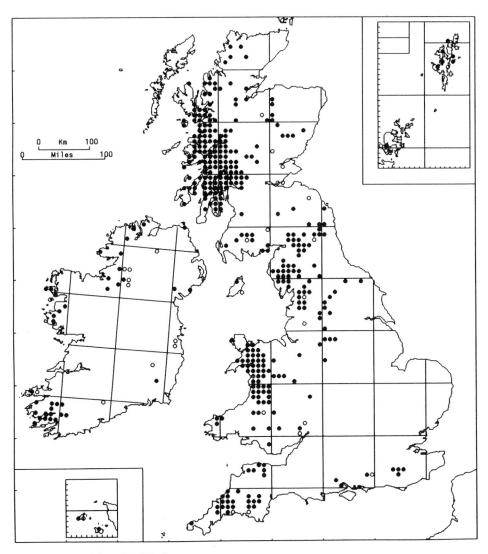

9/1. **Bazzania trilobata** (L.) S.F. Gray

This species is a calcifuge, particularly characteristic of western oak woodland where in humid conditions it can be locally dominant in the bryophyte layer, forming large, deep tufts on banks, boulders and outcrops. *Plagiochila spinulosa* and *Scapania gracilis* are common associates. It also grows on deep humus and brown-earth soils, especially in grazed woodland, as well as on logs; it is occasionally arboreal. In S.E. England, it is mainly associated with sandstone rocks in shaded valleys. Outside woodland, it occurs in sheltered places on boulder-strewn slopes and block-scree. Mostly below 500 m but reaching 900 m in the Cairngorms (above Loch Avon). GB 344+22*, IR 37+11*.

Dioecious; sporophytes rare.

Suboceanic in Europe; most frequent in the west, becoming uncommon in the extreme northern and continental regions and absent from most of the Mediterranean region. Azores, Madeira, E. Asia, Alaska, eastern N. America.

T. H. BLACKSTOCK

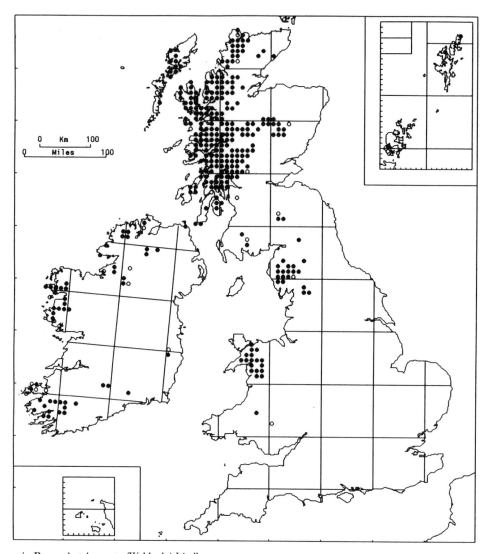

9/2. Bazzania tricrenata (Wahlenb.) Lindb.

Commonly occurs in upland dwarf-shrub heaths and on rocky banks, especially on N.- or E.-facing slopes, where it often grows in mixture with other bryophytes such as *Anastrepta orcadensis* and *Scapania gracilis*. Other habitats include cliff ledges, stable block-screes and submontane grassland; in addition it occurs among boulders and outcrops in western oak and birch woodlands. Most localities are base-poor. It ascends from sea-level in the west to 1175 m on Ben Lawers. GB 264+12*, IR 50+10*.

Dioecious; sporophytes not known in the British Isles.

Circumboreal. Boreal-montane in Europe, where widespread in Scandinavia but otherwise largely confined to the major mountain ranges from the Pyrenees eastwards to the Caucasus.

T. H. Blackstock

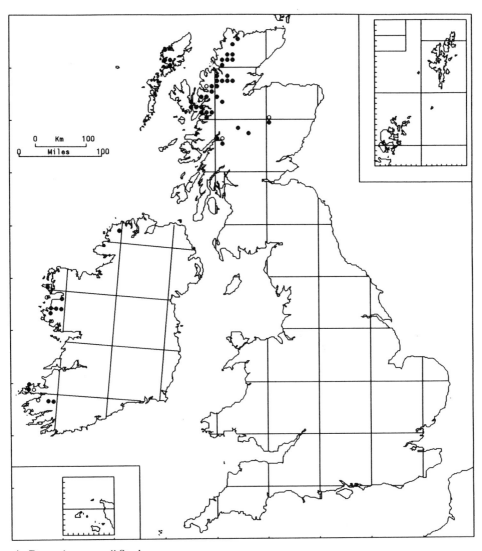

9/3. **Bazzania pearsonii** Steph.

A montane and strongly oceanic species, it is locally frequent in the N.W. Highlands of Scotland. It is a component of a distinctive mixed hepatic association with other large upland species, e.g. *Mastigophora woodsii* and *Scapania ornithopodioides*, in oligotrophic dwarf-shrub communities on moist peaty banks; it also occurs in acid grassland, block-screes, cliffs and ledges. Most localities have a northerly to easterly aspect and all have a cool and extremely humid climate, except in the Cairngorms where it is confined to areas of late snow-lie. 300–1000 m. GB 38+3*, IR 11+3*.

Dioecious; female plants occur in Britain, but neither males nor sporophytes have been reported.

Unknown in Europe outside the British Isles, this highly disjunct species has been recorded in E. and S.E. Asia and in north-west N. America.

T. H. BLACKSTOCK

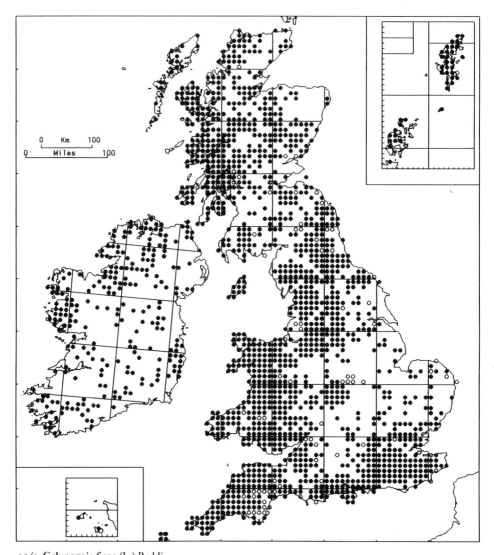

10/1. **Calypogeia fissa** (L.) Raddi

A common species forming small to extensive patches on more or less acidic, mineral or peaty soil in woods, moist turf, on heaths and banks, by streams, ditches and woodland tracks, on sandstone and acidic rock, decorticated logs, tussocks in bogs and fens, and growing through sphagnum. 0–850 m (Glen Coe). GB 1426+79*, IR 241+1*.

Autoecious or paroecious; sporophytes occasional, late winter or spring. Gemmae very common.

Throughout Europe from the Mediterranean region to southern Fennoscandia and S.W. Russia. Caucasus, Turkey, W. Himalaya, N. Africa, Macaronesia, eastern N. America.

A. J. E. Smith

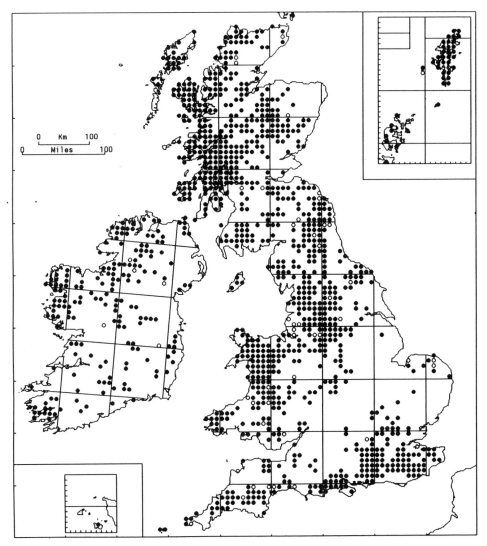

10/2. **Calypogeia muelleriana** (Schiffn.) K. Müll.

Of similar habit and in similar habitats to, but on more acid soils than, *C. fissa*. 0–900 m (bogs above Loch Einich, Cairngorms). GB 970+46*, IR 181+5*.

Autoecious or paroecious; sporophytes rare, April. Gemmae frequent.

Europe north to Fennoscandia and Russia but rare in the south and east. Caucasus, Macaronesia, N. America, Greenland.

A. J. E. SMITH

55

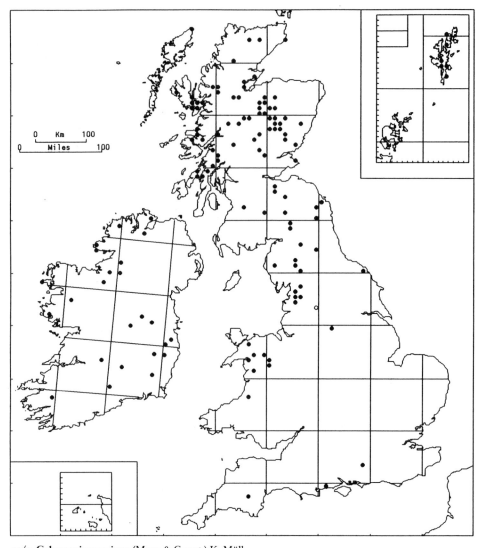

10/3. **Calypogeia neesiana** (Mass. & Carest.) K. Müll.

A strongly calcifuge plant of eroded peat in blanket bog, especially where there is some overhang; also in peaty ditches, on peaty banks and on humus-covered boulders and rotting logs, often with *C. muelleriana* and other small hepatics, rarely growing over *Leucobryum* or on mineral-soil banks in woodland. 0–760 m (Glas Tulaichean). GB 105 + 1*, IR 24.

Autoecious or paroecious; sporophytes rare, May and June. Gemmae rare.

Scattered localities in Europe from Spain, Italy and Yugoslavia north to Fennoscandia and Russia. Azores, Japan, Sakhalin, N. America, Greenland.

A. J. E. SMITH

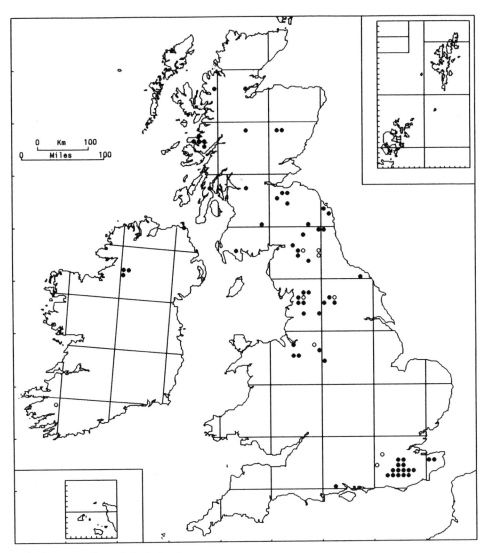

10/4. **Calypogeia integristipula** Steph.

This species most often occurs on sandstone outcrops and boulders but also grows on gritstones and sandy or peaty banks; usually in woodland but also in shaded treeless sites ranging from coastal slopes to montane block screes. It sometimes forms extensive mats with *C. muelleriana*. 0–850 m (S. Aberdeen). GB 61+8*, IR 4+1*.

Monoecious; plants occasionally fertile, but capsules very rare. Gemmae usually present.

Widespread in C. and N. Europe. Azores, Madeira, Caucasus, Japan, N. America, Greenland.

M. O. HILL

57

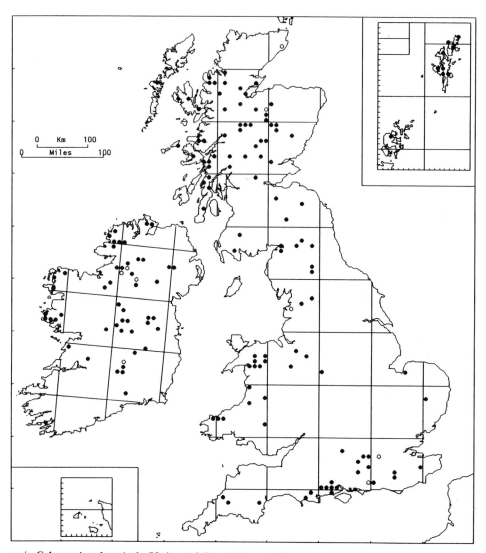

10/5. **Calypogeia sphagnicola** (H. Arn. & J. Perss.) Warnst. & Loeske

A characteristic plant of well-developed, saturated oligotrophic bogs, where it grows both among hummock-forming *Sphagnum* species such as *S. capillifolium* and *S. papillosum*, and to a lesser extent at the edge of hollows with *S. auriculatum*, *S. cuspidatum* and *Drepanocladus fluitans*. It is usually associated with other bog liverworts, e.g. *Cephalozia* spp., *Kurzia pauciflora* and *Mylia anomala*. 0–600 m (Breadalbane). GB 113+5*, IR 49+4*.

Monoecious; capsules rare, ripe May. Gemmae usually present.

Europe except the Mediterranean region. Macaronesia, Turkey, Japan, N. America, Greenland, Tierra del Fuego, Tasmania, New Zealand.

M. O. Hill

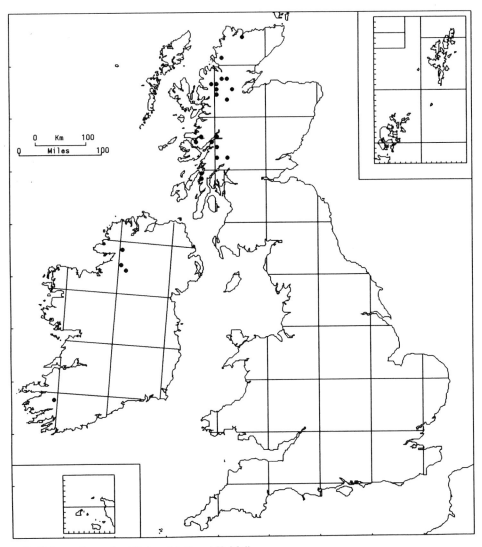

10/6. **Calypogeia suecica** (H. Arn. & J. Perss.) K. Müll.

This species is almost restricted to moist decorticated coniferous logs in very humid, wooded valleys. The plants, which are very small, occasionally form dense mats, but are more often scattered amongst other small liverworts characteristic of decaying logs; associated species sometimes include *C. muelleriana* and *C. fissa*. The two records from Co. Fermanagh are from relatively exposed peat; *C. suecica* may possibly have been overlooked in this habitat. 0–250 m. GB 19, IR 4.

Most populations appear to be dioecious but, in a gathering from Argyll bearing sporophytes, at least some of the plants were autoecious. Capsules ripe in spring. Attenuated gemmiferous shoots usually present.

Widespread in C. and N. Europe. Azores, Canaries, N. America.

M. O. Hill

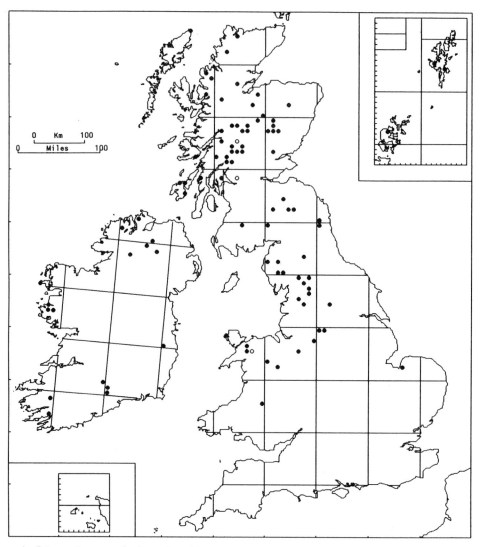

10/7. **Calypogeia azurea** Stotler & Crotz (*C. trichomanis* auct.)

Characteristically a calcifuge plant of peat, peaty or sandy soil, in blanket bog, on heaths, moorland and rocky banks; rarely, and possibly as an adventive, in woodland habitats. 0–800 m (Coire an t-Sneachda, Cairngorms). GB 76+3*, IR 16.

Autoecious or paroecious; sporophytes very rare, only immature found. Gemmae rare.

Montane and Arctic Europe from Spain, Portugal, Italy and Yugoslavia to N. Russia and Caucasus. Yenisei, Japan, Azores, Madeira, Tenerife, eastern N. America.

Confused in the past with other *Calypogeia* species; when the blue oil-bodies have been lost, some forms may be difficult or impossible to separate from *C. fissa* or *C. muelleriana.*

A. J. E. SMITH

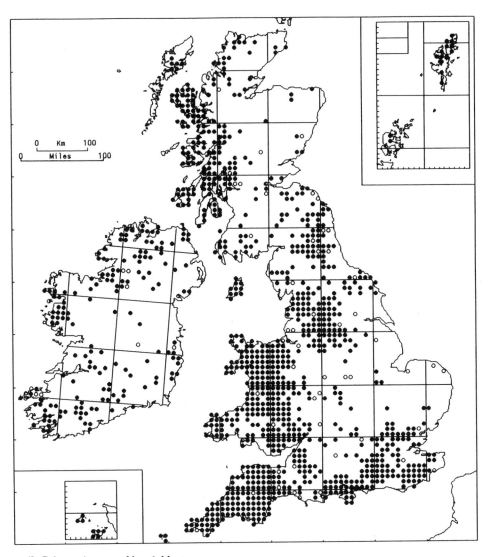

10/8. **Calypogeia arguta** Nees & Mont.

A calcifuge species of damp, especially clayey soil banks in sheltered habitats in woods, by paths and by streams, also at base of grass tussocks. Associated species often include *C. fissa* and *Dicranella heteromalla*. Mainly at low altitude, to 600 m in Wales (Plynlimon) and 610 m in S.W. Ireland (Ballaghbeama Gap). GB 923+56*, IR 160+12*.

Dioecious; capsules very rare, early spring. Gemmae very common.

Oceanic and Mediterranean Europe extending north to S.W. Norway and S. Sweden. Caucasus to Taiwan and Japan, Java, New Guinea, Africa, Macaronesia. A glasshouse adventive in N. America.

A. J. E. Smith

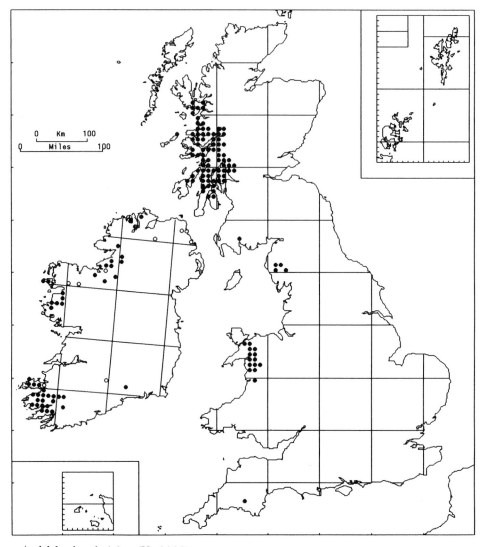

11/1. **Adelanthus decipiens** (Hook.) Mitt.

On dry but shaded acid or mildly basic blocks and low rock outcrops in light to moderate shade within low-lying natural or semi-natural deciduous woodlands, generally on E., S.E. or W. aspects. It also favours shaded blocks on N.-facing slopes where tree cover is open or absent. More rarely, it occurs on shaded rocks near the sea and on trees in extremely humid sites such as within the spray zone of waterfalls. It is commonly associated with *Hymenophyllum wilsonii*, *Bazzania trilobata*, *Lepidozia cupressina*, *Scapania gracilis*, *Plagiochila spinulosa* and *Saccogyna viticulosa*. Most plentiful and luxuriant at low elevations (0–200 m) but in Co. Kerry it reaches 610 m; with distance northwards and eastwards it becomes confined to increasingly lower elevations and all Scottish localities appear to lie below 200 m. GB 98, IR 43+11*.

Dioecious; British and Irish plants all male.

Occurs in W. France, Spain. Macaronesia, tropical and southern Africa, Caribbean, St Helena and S. America. The British and Irish localities are its northernmost world stations.

H. J. B. Birks & D. A. Ratcliffe

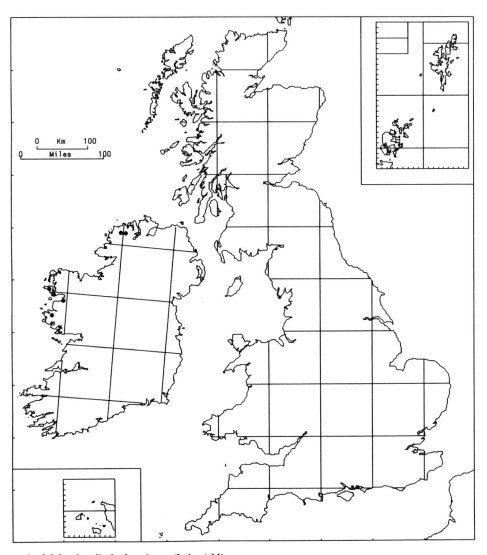

11/2. **Adelanthus lindenbergianus** (Lehm.) Mitt.

Locally frequent in open, oligotrophic heather or other dwarf-shrub communities on acid mountains. It favours steep blocky slopes on N.-, N.E.- or E.-facing slopes and occurs commonly mixed with *Bazzania tricrenata*, *Herbertus aduncus*, *Mylia taylorii*, *Plagiochila spinulosa*, *Pleurozia purpurea*, *Scapania gracilis*, *Breutelia chrysocoma* and *Racomitrium lanuginosum*. It also occurs on broken cliffs, block-screes, cliff ledges and, more rarely, in damp but well-drained grassland below cliffs but always on shaded aspects. Mainly at elevations 500–800 m. IR 6.

Dioecious; Irish plants female.

Absent elsewhere in Europe. Recorded from tropical Africa, southern Africa, C. and S. America, Antarctica. In view of its habitat preferences, its absence from Co. Kerry and W. Scotland is surprising.

H. J. B. Birks

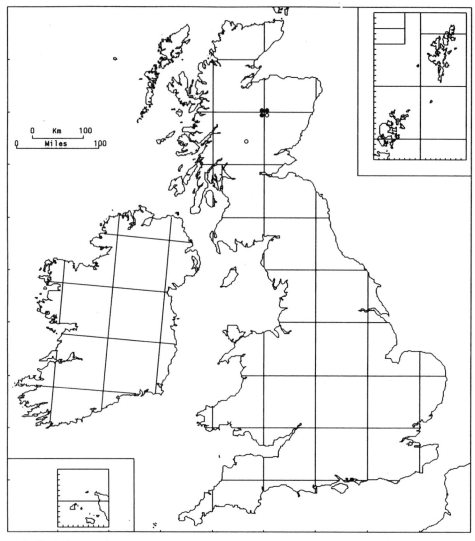

12/1. **Cephalozia ambigua** Mass. (*C. bicuspidata* (L.) Dum. ssp. *ambigua* (Mass.) Schust.)

On bare, moist soil on high mountains. GB 3+2*.
   Monoecious.
   N. Europe and mountains of C. Europe. N. America, Greenland.
   An obscure and much disputed taxon, treated by most recent authors as a cytological subspecies of *C. bicuspidata* with n=9. The degree to which cytological differences are reflected in morphology is still uncertain. Plants mapped here have been determined morphologically rather than cytologically.

M. O. HILL

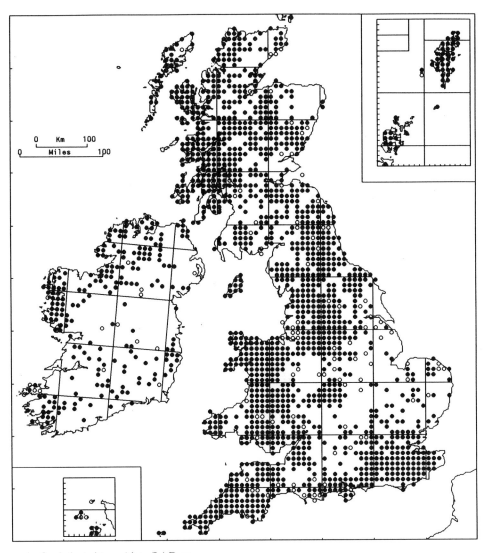

12/2. **Cephalozia bicuspidata** (L.) Dum.

The ecological range of this very common calcifuge overlaps that of the other *Cephalozia* spp. in the British Isles. Equally tolerant of moist mineral and organic soils, as well as growing on rotting logs and damp rock exposures, it occurs in woodlands and in many open habitats, ascending into the montane zone. It also grows in sphagnum bogs and on tussocks in fens, and persists in partially drained mires. In addition, it is invasive in disused quarries, road-cuttings, ditch-sides and other highly modified sites. 0–1170 m (Aonach Beag). GB 1473+99*, IR 211+16*.

Monoecious; sporophytes frequent. Reports of gemma production uncommon.

Circumboreal. Throughout Europe, but becoming infrequent in the Mediterranean region.

In Europe, *Cephalozia bicuspidata sens. lat.* includes three chromosome races. *C. ambigua* is mapped separately.

T. H. BLACKSTOCK

65

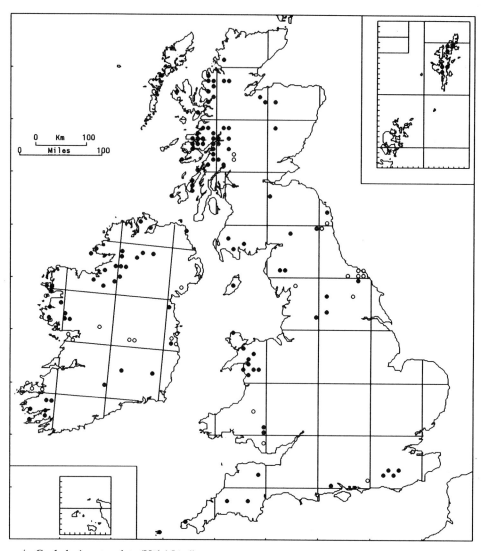

12/3. **Cephalozia catenulata** (Hüb.) Lindb.

A characteristic though usually infrequent member of the bryophyte community that develops on damp, rotting tree-trunks in humid woodlands in the west. Associates include *Cephalozia lunulifolia*, *Nowellia curvifolia*, *Riccardia palmata* and *Scapania umbrosa*. It also grows on steep, shaded rock-faces, for example on moist sandstone outcrops in valleys in the Sussex Weald. Outside woodland it occurs on peaty soil and banks, often in sheltered rocky ground. Predominantly lowland, up to 500 m (E. Scotland). GB 90+15*, IR 38+14*.

Dioecious; sporophytes uncommon. Gemma-bearing plants occur but are seldom reported so that their frequency cannot be properly assessed.

Circumboreal. Europe north to S. Scandinavia and W. Russia, montane except near its northern limit.

T. H. BLACKSTOCK

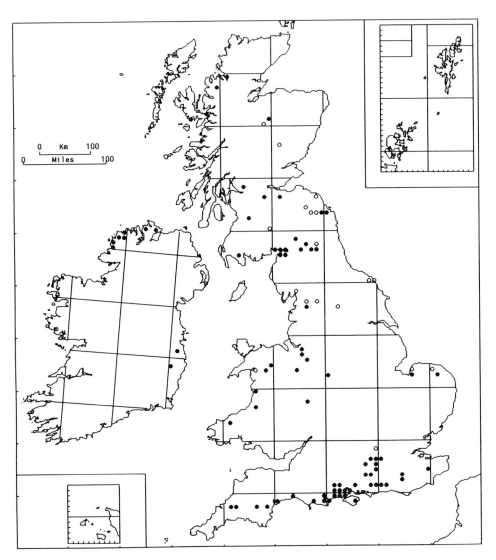

12/4. **Cephalozia macrostachya** Kaal.

This is a bog hepatic, largely confined to saturated micro-habitats on ombrogenous mires, valley bogs, acid basin-mires and very wet heathland. It generally grows amongst sphagnum, and it often associated with other liverworts, including *C. connivens*, *Cladopodiella fluitans*, *Kurzia pauciflora* and *Mylia anomala*. Lowland. GB 71+19*, IR 9.

Dioecious; sporophytes rare. Gemmae sometimes reported.

In N. and C. Europe, where mainly restricted to peatlands. Eastern N. America.

Fertile material of *C. macrostachya* is normally required for reliable determination and, although some records of sterile plants may be mapped, it is probably somewhat under-recorded overall.

T. H. BLACKSTOCK

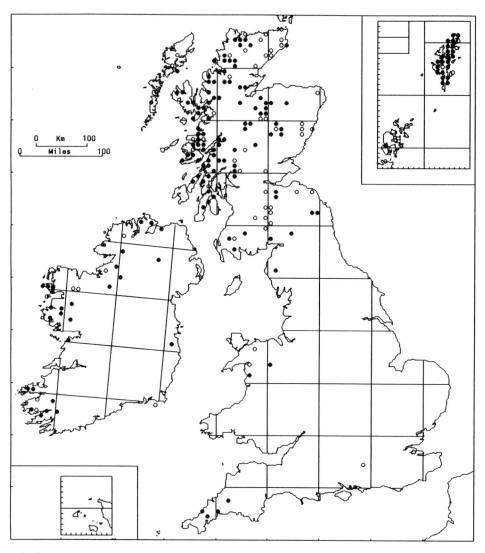

12/5. **Cephalozia leucantha** Spruce

A submontane species, found mainly on wet heath and moorland, and occasionally in peat-bogs. It usually grows on damp peat and peaty soil on sheltered banks, streamsides and ditch edges, or amongst other bryophytes, especially sphagnum. It is less frequent on rotting logs and stumps. 0–600 m (Lanarkshire). GB 126+52*, IR 28+9*.

Dioecious; sporophytes uncommon but have been recorded from Cornwall to Caithness. No information about occurrence of gemmae in Britain but reported from Europe and N. America.

Circumboreal. Boreal-montane in N.W. and C. Europe.

Appears to be genuinely very rare in most of England and Wales; the paucity of modern records in the Southern Uplands of Scotland may reflect lack of recent survey and not necessarily a decline.

T. H. BLACKSTOCK

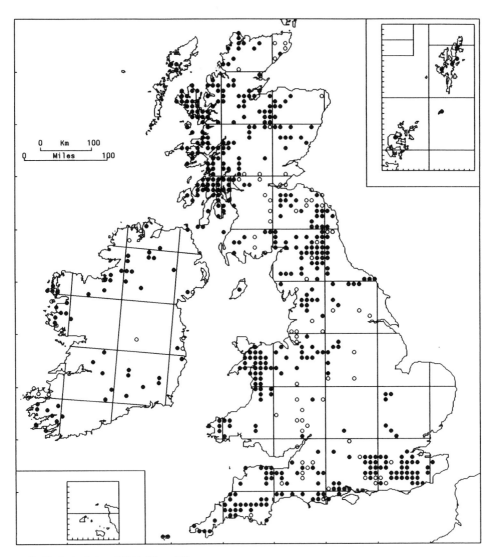

12/6. **Cephalozia lunulifolia** (Dum.) Dum.

Found in a variety of acidic habitats, but most frequent in deciduous woodland where it grows on rotting tree-trunks and stumps, with associates such as *Lepidozia reptans* and *Tetraphis pellucida*, around tree-bases, on humus-banks and, occasionally, on rock outcrops. Elsewhere, occurs on thin peaty soil in sheltered positions on rocky slopes and screes. Also a species of acid mires, wet moors and damp heaths, found amongst sphagnum, on peat and by ditches. Mostly in the lowlands, reaching 600 m in the Southern Uplands of Scotland. GB 456+74*, IR 53+11*.

Dioecious; sporophytes occasional and recorded in many parts of the British Isles. Gemmae frequent.

Circumboreal. Widely distributed in Europe, in montane, boreal and broadleaved forest regions.

T. H. BLACKSTOCK

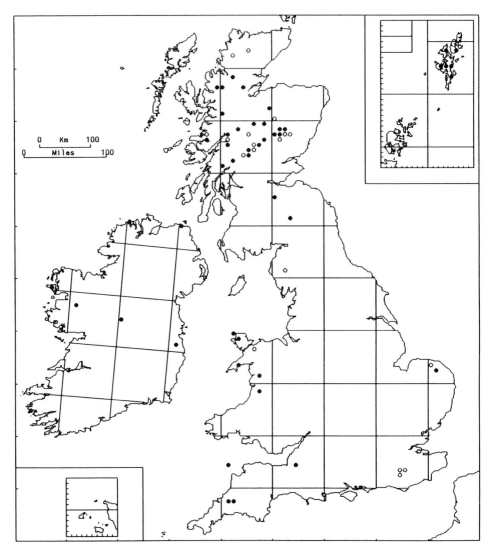

12/7. **Cephalozia pleniceps** (Aust.) Lindb.

The lax wetland form of this species has been recorded throughout its range, in peat-bogs, acid flushes and wet heathland, as well as by streams and on tussocks in fens. In these habitats, which are mostly in the lowlands, it typically grows amongst sphagnum with associates such as *C. connivens*, *Mylia anomala* and *Riccardia latifrons*. In the Scottish Highlands it also occurs at higher altitudes in more compact tufts on moist ledges and banks, sometimes on mineral soil which can be moderately base-rich. 0–1080 m (Aonach Beag). GB 40+18*, IR 7.

Autoecious; usually fertile and often found with sporophytes. Gemmiferous shoots have been observed in Scotland but their frequency is unknown.

Circumboreal. Widespread in N. Europe and in montane regions further south.

Somewhat under-recorded, especially in Ireland where it was not detected until 1968.

T. H. BLACKSTOCK

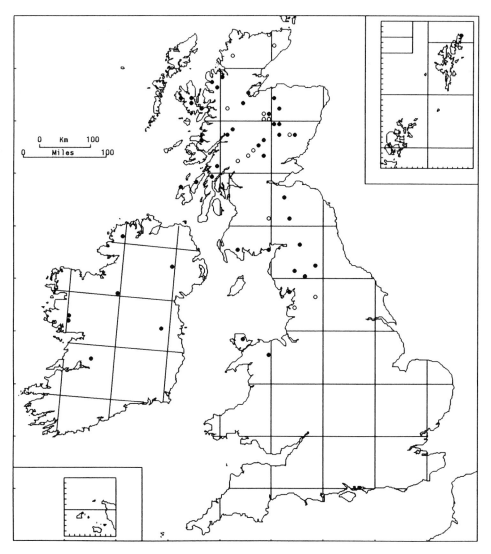

12/8. **Cephalozia loitlesbergeri** Schiffn.

This species, which was first described in 1912, occurs on raised bogs, blanket mires, valley bogs and wet moors, and also occasionally in damp heathland. It usually grows amongst *Sphagnum*, including *S. imbricatum* ssp. *austinii* and *S. magellanicum* on ombrogenous mires, sometimes in mixture with other bog hepatics such as *Calypogeia sphagnicola*, *Mylia anomala* and *Odontoschisma sphagni*. Mainly lowland but recorded up to 700 m. GB 38+14*, IR 7.

    Autoecious; sporophytes have been recorded and may be fairly frequent.

    Principally in the boreal and montane regions of N. and C. Europe. Eastern N. America.

T. H. BLACKSTOCK

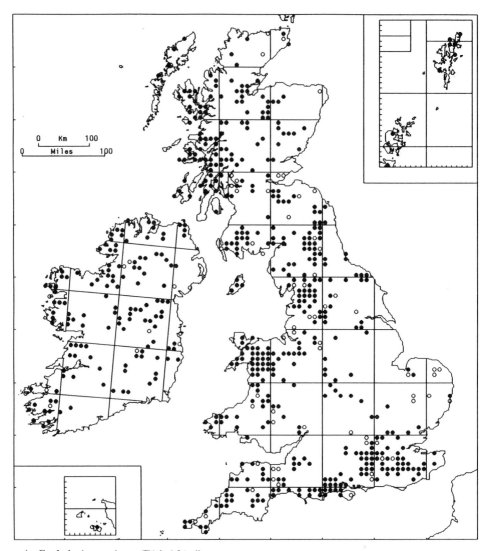

12/9. **Cephalozia connivens** (Dicks.) Lindb.

Widespread at low altitudes in ombrogenous bogs, valley mires, damp heathland and moorland. It often grows among sphagnum, sometimes with other *Cephalozia* spp., but is also found in tussocks and on moist peaty soil. In saturated peatlands, it commonly occurs with other bog hepatics, e.g. *Kurzia pauciflora* and *Mylia anomala*. Less frequently it is found on old stumps and decaying logs in woodland, and very occasionally on shaded rock-faces. Mainly lowland but recorded at 650 m on Ben Vorlich, Dunbartonshire. GB 449+62*, IR 132+12*.

Monoecious; sporophytes frequent. Gemmae rare. Like *Kurzia* spp., can regenerate from buried underground axes on disturbed peat bogs (Duckett & Clymo, 1988).

Circumboreal. Widespread in Europe, becoming rare in the far north and montane in the south.

T. H. BLACKSTOCK

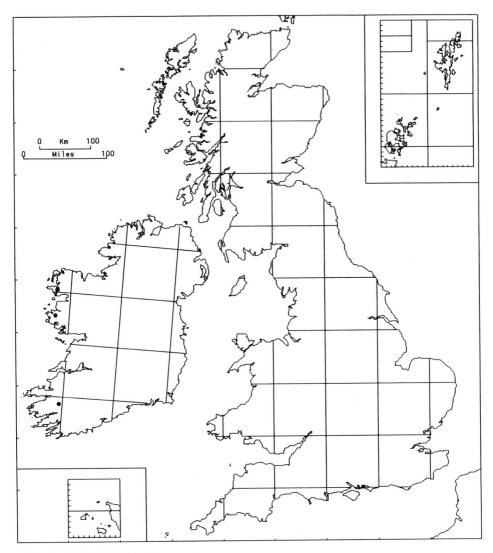

12/10. **Cephalozia hibernica** Spruce ex Pears.

Since it was first collected in 1865, this species has been found in a handful of localities near Killarney. More recently it has been detected in one area of W. Galway (1968) and on Achill Island (1987). It occurs on steep rock-faces, boulders, humus-banks and exposed roots, often under heavy shade, in deciduous woodlands, wooded ravines and rhododendron thickets. Lowland. IR 3.

Dioecious; female and male shoots have been described from Killarney material, but sporophytes have not been reported. Irish plants produce gemmae.

An Atlantic hepatic hitherto known only from Ireland, Spain, Madeira, Azores. However, according to Váňa (1988), *C. hibernica* is a synonym of *C. crassifolia* (Lindenb. & Gott.) Fulf., which is known from S. America, the Caribbean and Mexico.

T. H. BLACKSTOCK

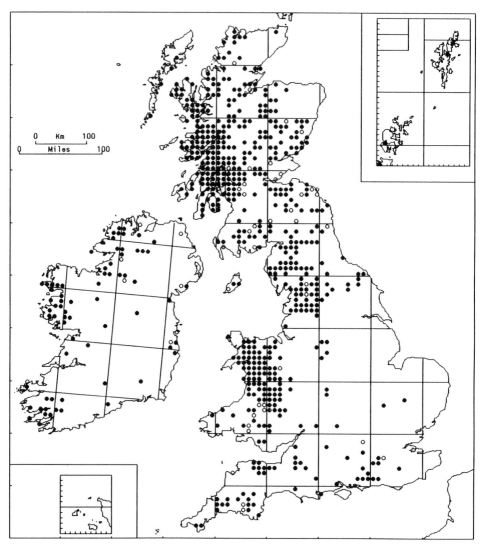

13/1. **Nowellia curvifolia** (Dicks.) Mitt.

A species occurring widely on rotting timber, most commonly in deciduous woodland. Stumps and the decorticated trunks and branches of a range of broadleaved trees and conifers are colonized. It is often forms extensive patches, sometimes in mixture with other hepatics, including various *Cephalozia* spp., *Lophocolea heterophylla* and *Odontoschisma denudatum*. It also occurs on peaty soil, frequently in moorland, heathland and dry peat-bogs, chiefly in the high-rainfall areas of the north and west. Very rarely, it is rupestral. Mostly lowland, but reaching 800 m in W. Scotland (Glen Duror). GB 539+42*, IR 70+9*.

Autoecious or dioecious; sporophytes frequent and widespread. Gemmiferous shoots produced on wood and peat, and may be frequent but not recorded consistently.

Suboceanic in Europe, very infrequent in the extreme north and the Mediterranean region. Macaronesia, E. and tropical Asia, eastern N. America, C. America.

N. curvifolia is evidently spreading in the south of England and Wales, where it was virtually unknown prior to 1950. Recent increases have also been reported in Belgium and adjoining territories, particularly in spruce plantations in the Ardennes.

T. H. BLACKSTOCK

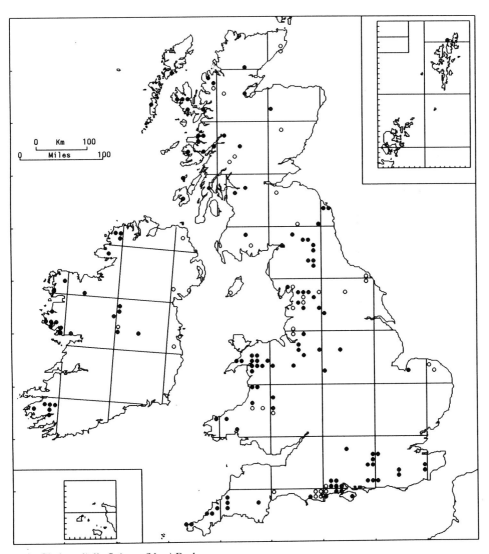

14/1. **Cladopodiella fluitans** (Nees) Buch

This is a species of saturated peatlands, found in ombrogenous bogs, acidic valley and basin mires, and occasionally in wet heaths and moors. Unlike most other bog hepatics, it is sometimes fully submerged, often with *Sphagnum cuspidatum*, in shallow bog pools; but it also occurs on peat and amongst other *Sphagnum* spp., e.g. *S. magellanicum*, *S. papillosum* and *S. recurvum*, on hummocks and lawns, frequently associated with liverworts such as *Cephalozia connivens*, *Kurzia pauciflora* and *Odontoschisma sphagni*. Predominantly lowland, ascending to about 600 m. GB 108+35*, IR 26+6*.

Dioecious; sporophytes rare.

Circumboreal. Widely distributed in N. and C. Europe.

T. H. BLACKSTOCK

75

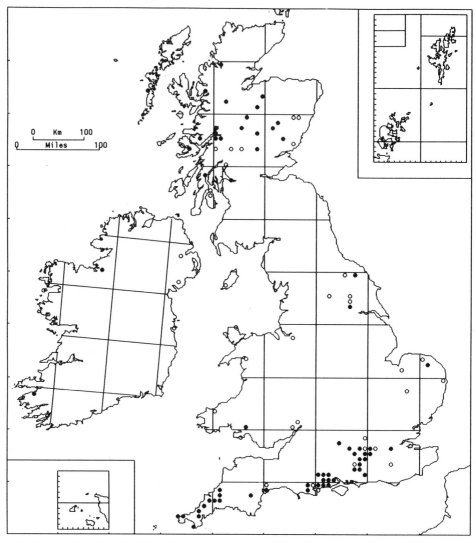

14/2. **Cladopodiella francisci** (Hook.) Jørg.

Although generally uncommon, this species has been widely recorded in areas of damp lowland heathland in the south of England, where it can be locally abundant. Further north it is also found in bogs, moorland and lakesides, and forms patches on moist peaty soil on banks and ditch edges. Occasionally it occurs on damp sandy soil or gravel, on compacted soil on footpaths, and on soil-capped walls. Mostly lowland, but to 1000 m on Lochnagar. GB 63+33*, IR 2+5*.

Dioecious; sporophytes very rare but recorded in the southern and northern parts of its range. Gemmae frequent.

N. Europe and mountains of C. and W. Europe, south to Azores and Madeira. Contrary to what one would expect from its British distribution, it is chiefly montane in Fennoscandia, extending to the Arctic. Eastern N. America, Greenland.

T. H. BLACKSTOCK

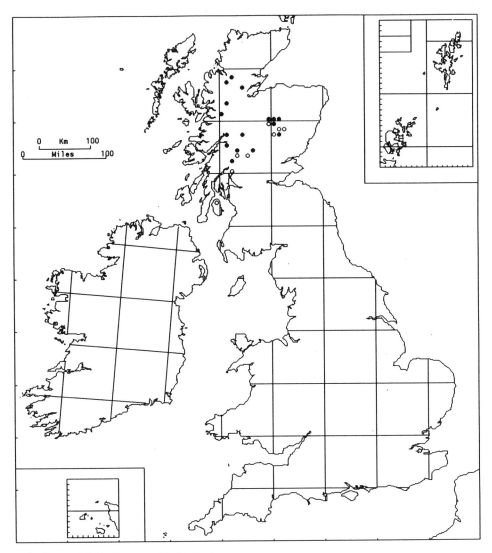

15/1. **Pleurocladula albescens** (Hook.) Grolle

A species of late snow-bed communities on some of the Scottish mountains. It grows in pure tufts or amongst other bryophytes on moist soil near the edges of snow-patches and in gullies and boulder-scree irrigated by melt-water. Associates include more widespread species, e.g. *Gymnomitrion concinnatum* and *Polytrichum alpinum*, as well as others, e.g. *Moerckia blyttii*, similarly restricted. In some circumstances deep snow-cover can be greatly prolonged, as witnessed in 1956 when *P. albescens* was recorded on 9 August under 1.4 m of snow in the Cairngorms. Mostly above 900 m, extending to 1335 m on Ben Nevis. GB 16+8*.

Dioecious; sporophytes unknown in Britain. Gemmae have been noted in material collected on Lochnagar.
Circumpolar. Arctic-alpine; in the far north and the mountains of C. Europe.

In recent years British bryologists have taken a broad view of variation within *P. albescens*, and slender forms, approaching '*P. islandica*', have not been given taxonomic recognition.

T. H. BLACKSTOCK

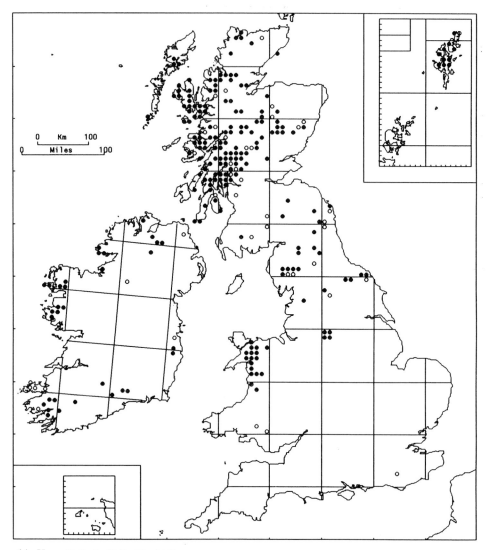

16/1. **Hygrobiella laxifolia** (Hook.) Spruce

It forms appressed patches on wet rocks beside streams and on flushed rock outcrops and crags, frequently growing with *Blindia acuta* and *Jungermannia* spp. It is most often associated with rocks that are at least mildly basic, including basalts and schists, although also occurring on more acidic substrates. In Sussex it grew on sandstone rocks in a stream, but has not been seen for 70 years. In Shetland and Skye it has been noted from moist silty or sandy ground on loch margins, growing with weedy species such as *Archidium alternifolium*. 0–1070 m (Aonach Beag). GB 180+35*, IR 32+9*.

Dioecious; female plants and sterile perianths abundant, male plants apparently rarer. Capsules uncommon, maturing in summer. Gemmae unknown.

N. Europe and mountains of C. and W. Europe. Japan, N. America, Greenland.

M. M. YEO

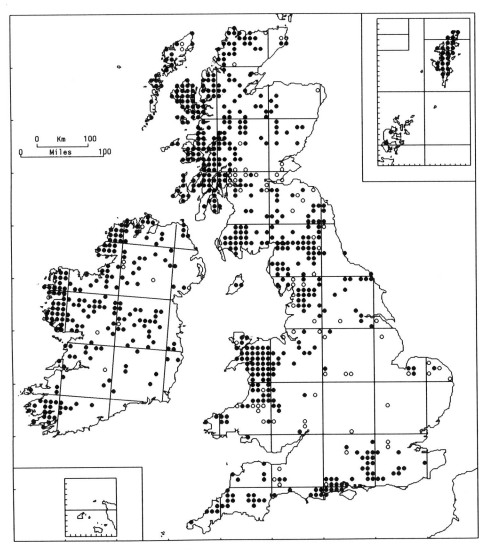

17/1. **Odontoschisma sphagni** (Dicks.) Dum.

A characteristic species of ombrogenous mires and valley bogs, and also widespread in moorland and damp heathland. It most frequently grows amongst *Sphagnum* spp., of which *S. papillosum* is a particularly common associate, but can form patches on wet peat, the base of tussocks and decaying vegetation. In saturated peatlands *O. sphagni* is sometimes locally prominent amongst other bog hepatics, including *Cephalozia* spp., *Kurzia pauciflora* and *Mylia anomala*. Mostly lowland, but reaching 880 m (Coire an Lochain, Cairngorms). GB 544+76*, IR 207+9*.

Dioecious; sporophytes are infrequent.

Suboceanic in Europe, including Iceland and the Faeroes. Also reported from the Azores and Madeira, with a very restricted distribution in eastern N. America.

T. H. BLACKSTOCK

79

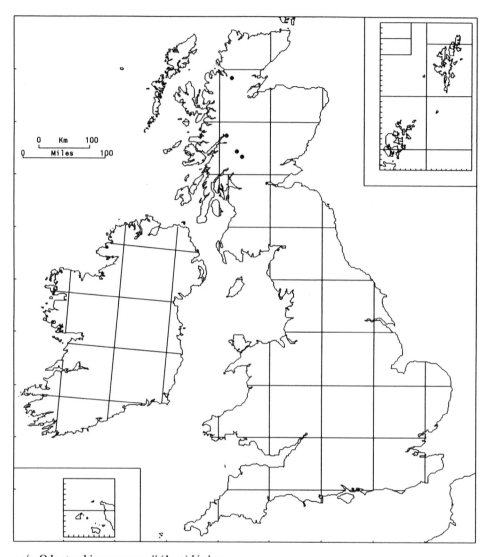

17/2. **Odontoschisma macounii** (Aust.) Underw.

On thin moist micaceous soil over friable base-rich schist in rock crevices and on ledges and slabs, often in sheltered gullies or on N.-facing cliffs. Associated species include *Jungermannia subelliptica*, *Amphidium mougeotii*, *Anoectangium warburgii*, *Ctenidium molluscum* and *Distichium capillaceum*. Restricted in Britain to a few Scottish mountains, in the altitudinal range 750–880 m (Beinn Heasgarnich). GB 4.

Dioecious; Scottish plants are sterile and lack gemmae.

Circumpolar. An arctic-alpine; occurring in the Alps and northern Europe. It is common in the Arctic, where it is often abundant in calcareous mossy tundra.

First discovered in 1900 by P. Ewing on Beinn Heasgarnich in Perthshire where it still grows; most localities have been added since 1980.

D. G. Long

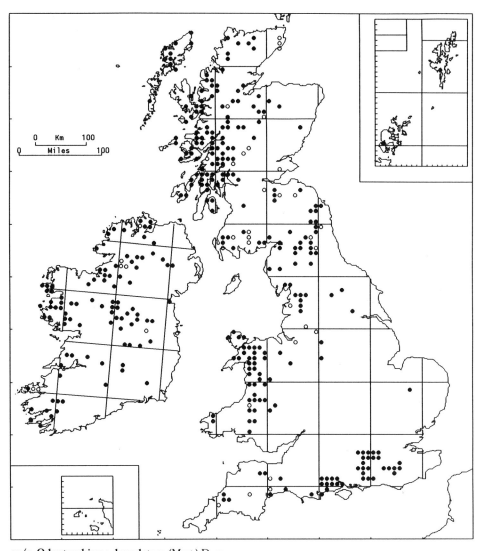

17/3. Odontoschisma denudatum (Mart.) Dum.

This is one of a number of liverworts found both on rotting logs and peaty soil. In the British Isles it most frequently occurs on the latter, in heathland and moorland, on rocky slopes and boulder-scree, commonly on rather shallow organic soils. It sometimes grows in bogs, especially in partially drained and cut-over sites, on damp but not saturated peat, thus hardly ever overlapping *O. sphagni*. It can also be locally abundant on rotting logs and stumps in deciduous woodland, and occasionally occurs in conifer plantations. Mainly lowland, up to 400 m (Widdybank Fell). GB 248+34*, IR 94+8*.

Dioecious; sporophytes are very rare. Gemmae extremely frequent, generally produced in abundance.

Widely distributed in W., N. and C. Europe. Himalaya, Japan, eastern N. America, Caribbean, C. and S. America.

T. H. BLACKSTOCK

81

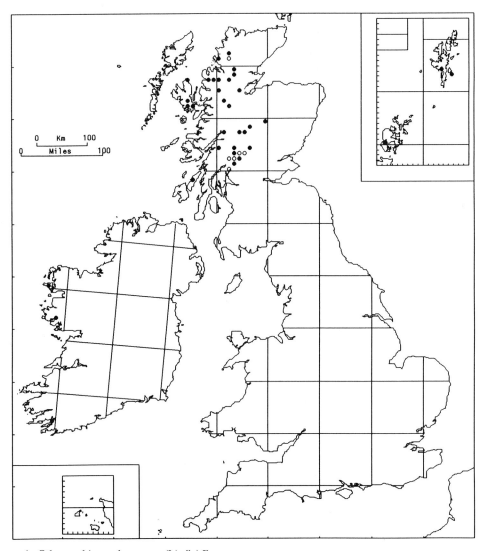

17/4. **Odontoschisma elongatum** (Lindb.) Evans

Typically found growing in stony or gravelly peaty flushes and on loch margins, often with *Blindia acuta*, *Calliergon sarmentosum* and *Drepanocladus revolvens*. Although frequently in quantity and in relatively pure, prostrate patches, its dark brown colour makes it easy to overlook in the field. In other habitats it is rarer, as on sandy sea-banks in Co. Mayo, amongst sphagnum in peaty hollows, and even on pure moist peat; in the last two habitats it overlaps ecologically with *O. sphagni*. In Scotland there is no clear altitudinal preference, records being evenly spread from 10 m (Loch Maree) to 910 m (Ben Lawers). GB 30+8*, IR 3.

Dioecious, mostly female; male plants and sporophytes very rare. Gemmae are rare.

Distribution in Europe mainly northern, with scattered localities in the central mountains. N. America, Greenland.

Although known from Scotland in the early part of this century, the majority of records are recent (post–1960). Easily overlooked, it is certainly less rare than the present data suggest; the extent of its distribution in the British Isles is only now becoming apparent. It was first collected in Ireland in 1968.

D. G. Long and T. H. Blackstock

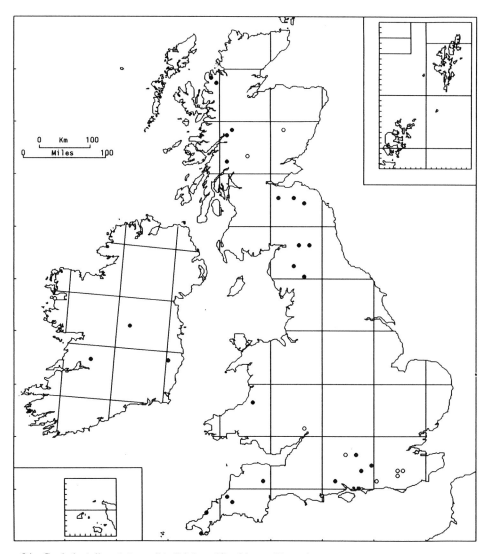

18/1. **Cephaloziella spinigera** (Lindb.) Jørg. (*C. subdentata* Warnst.)

Almost confined to bogs, where it is most frequently found growing amongst sphagnum or other mosses, notably *Leucobryum glaucum*, less often on peaty hummocks beneath *Calluna*. It has also been recorded from steep peaty banks. GB 24+9*, IR 5.

Autoecious; sporophytes frequent. Gemmae fairly frequent.

C. and N. Europe. N. America.

In the past, it has been confused with *C. elachista*. The map is based on material revised by J. A. Paton.

M. F. V. CORLEY

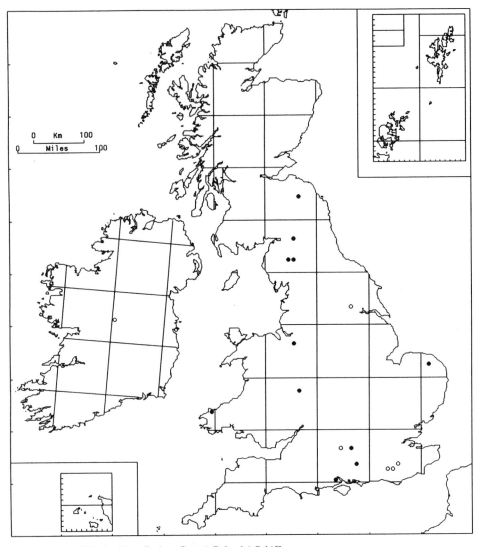

18/2. **Cephaloziella elachista** (Jack ex Gott. & Rabenh.) Schiffn.

Like many bog hepatics, this generally grows amongst larger bryophytes, especially sphagnum, in which it may be mixed with *Mylia anomala* and *Odontoschisma sphagni*, in raised and valley bogs. On wet heathland it has been found on the sides of decaying *Molinia* tussocks. Lowland. GB 11+5*, IR 2*.

Autoecious; perianths and sporophytes occur. Gemmae are frequently present.

Widespread but local in C. and N. Europe. N. America.

In the past, it has been much confused with *C. spinigera*. The map is based on material revised by J. A. Paton.

M. F. V. CORLEY

84

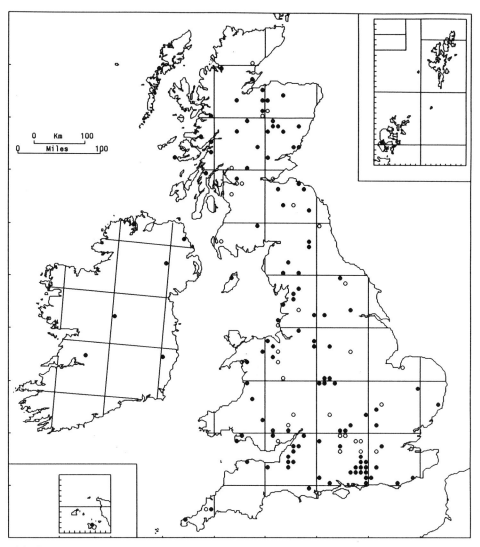

18/3. **Cephaloziella rubella** (Nees) Warnst.

Occupies a similar range of habitats to *C. hampeana*, but less characteristic of moist habitats, being scarcer in sphagnum bogs and more often found on well-drained sand and gravel. It is probably the most frequent *Cephaloziella* on decorticated logs and stumps. 0–900 m (Coire Cheap, Ben Alder range). GB 114+35*, IR 6.

Paroecious or occasionally autoecious; sporophytes frequent, summer. Gemmae frequent.

C. and N. Europe, becoming very scarce in the south, and not recorded from Portugal, Corsica or Greece. Macaronesia, N. Asia, N. America.

Many unnamable sterile gatherings probably belong to *C. rubella* or *C. hampeana*, and in this way the species is probably under-recorded.

M. F. V. CORLEY

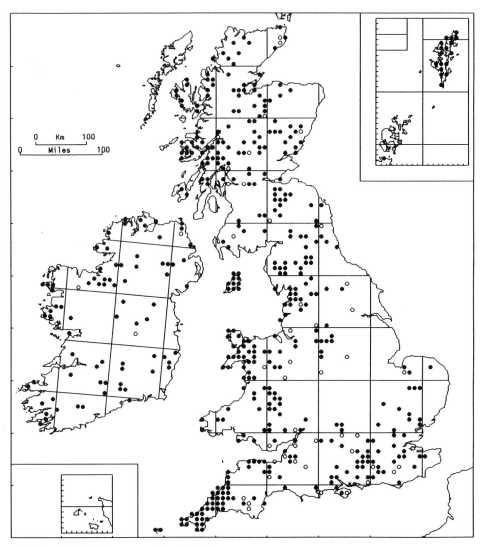

18/4. **Cephaloziella hampeana** (Nees) Schiffn.

Widespread in a range of acid habitats, growing on peaty heathland and moorland banks and among sphagnum in bogs, on clay in woodland rides, on sand and gravel in quarries, by tracks and on dunes, on mine waste, on soil among rocks, especially on sea-cliffs and on rock-faces in sheltered places such as block-screes and treeless ravines. It is absent from some of the drier sites that *C. divaricata* favours. 0–1175 m (Ben Lawers). GB 394+45*, IR 76+3*.

Autoecious; sporophytes common throughout the year. Gemmae frequent.

C. and N. Europe. Macaronesia, temperate Asia, N. America.

M. F. V. CORLEY

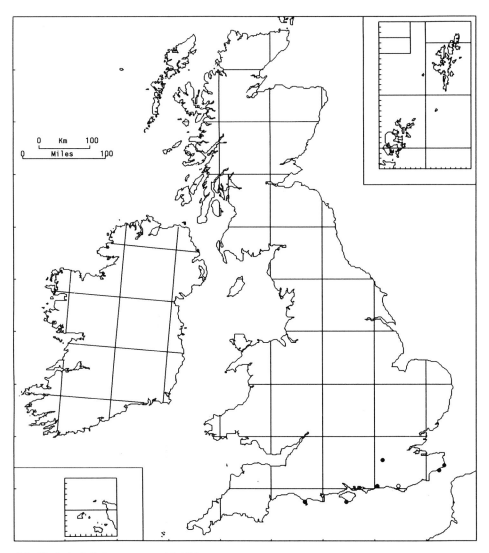

18/5. **Cephaloziella baumgartneri** Schiffn.

The only calcicolous *Cephaloziella*, growing on chalk, limestone and calcareous Greensand rocks on cliffs and in quarries, on stones partly buried in turf and on soil among rocks. Lowland. GB 6+1*.

Autoecious; sporophytes frequent. Gemmae common.

Mediterranean-Atlantic, extending from Turkey to the Canaries and northwards to Belgium and southern England.

M. F. V. CORLEY

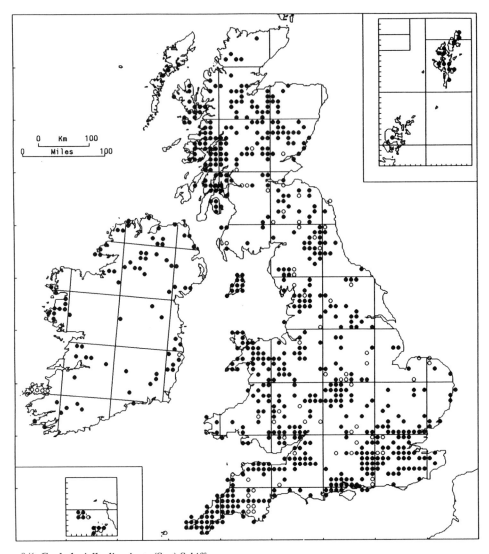

18/6. **Cephaloziella divaricata** (Sm.) Schiffn.

Generally found in open sites on acid substrates, this species avoids the wet habitats, such as sphagnum bogs, in which *C. hampeana* is often found. It grows on peaty, sandy and gravelly soils on sea-cliffs, heaths, mine spoil, tracks, wall-tops, quarries, fixed dunes, woodland rides, turf overlying rocks and in rock crevices; also occasionally in shady habitats, such as boulders in woods, tree boles and rotten wood. 0–990 m. GB 645 + 68*, IR 80 + 8*.

Dioecious; sporophytes occasional. Gemmae frequent. Sterile plants are common, and can be identified if non-gemmiferous shoots are present.

Throughout Europe and most of the temperate Northern Hemisphere.

M. F. V. CORLEY

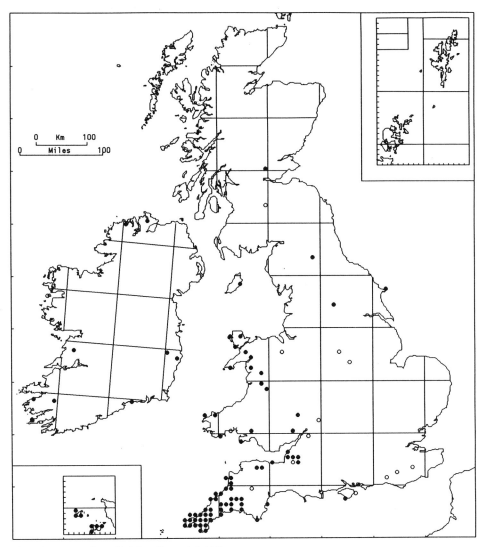

18/7. **Cephaloziella stellulifera** (Spruce) Schiffn.

Most records are from mine waste, or from coastal sites, where it grows on heathland, soil in rock crevices on cliffs and in turf on cliff-tops. It is also found occasionally on heathy tracks or sandy ditch-banks inland, especially in plantations. Lowland. GB 73+13*, IR 9+1*.

Paroecious, or sometimes autoecious; sporophytes frequent, spring. Gemmae frequent.

Principally southern and western in Europe, but reaching as far as Sweden, Poland and Hungary. Widespread in temperate parts of the Northern Hemisphere.

M. F. V. CORLEY

89

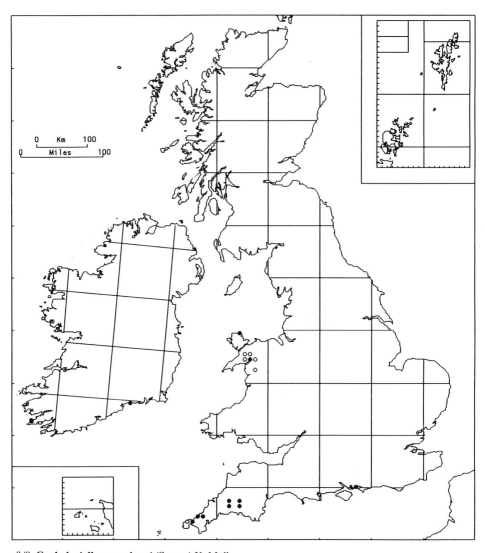

18/8. Cephaloziella massalongi (Spruce) K. Müll.

Confined to copper-bearing rocks and associated soils, it is chiefly found on stony slopes of mine waste or the silt washed from it, and in wall crevices around disused copper mines, in moist to wet situations. Associates include *Gymnocolea inflata* and *Pohlia nutans*, and also other *Cephaloziella* spp., notably *C. divaricata*, *C. nicholsonii* and *C. stellulifera*. 0–600 m (Snowdon). GB 9+6*, IR 2.

Allegedly autoecious, but possibly dioecious also; female inflorescences rare, male inflorescences and sporophytes unknown in Britain. Gemmae common.

W. Europe eastwards to Italy, Austria, Czechoslovakia, Sweden, Bulgaria. N. America.

The map is based on material revised by J. A. Paton (cf. Paton, 1984).

M. F. V. CORLEY

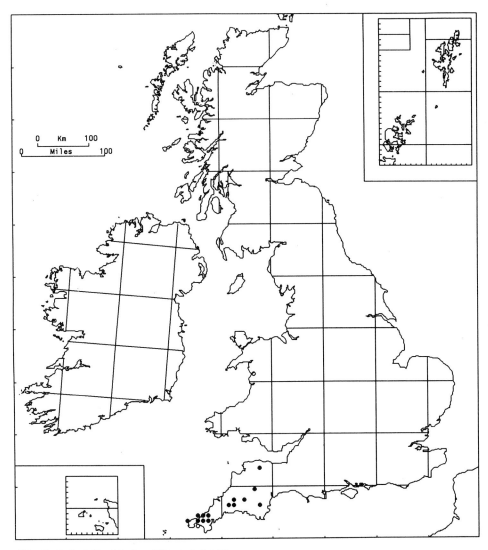

18/9. **Cephaloziella nicholsonii** Douin & Schiffn.

Found in the same situations as *C. massalongi* but able to tolerate rather drier conditions. Lowland. GB 14+2*.
  Paroecious; perianths are frequent but sporophytes are unknown. Gemmae common.
  Not definitely known from outside Britain.
  The map is based on material revised by J. A. Paton (cf. Paton, 1984).

M. F. V. CORLEY

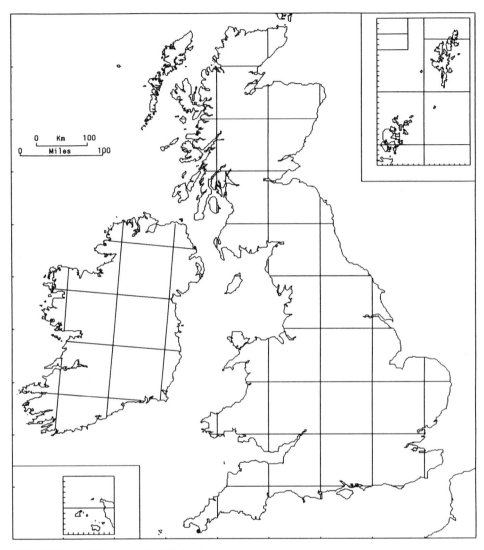

18/10. **Cephaloziella dentata** (Raddi) Steph.

A plant of depressed ground such as cart-tracks, where liable to periodic shallow flooding. Confined to the serpentine heathland of the Lizard peninsula, where it was first found in 1926. Lowland. GB 1.

Dioecious; perianths occur but male plants are unknown. Gemmae are usually present.

A Mediterranean-Atlantic species recorded from Madeira and extending eastwards to Yugoslavia and northwards to S.W. England (Cornwall) and Denmark.

M. F. V. CORLEY

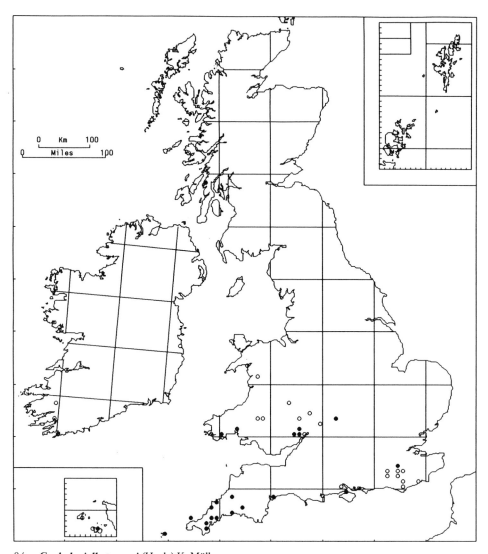

18/11. **Cephaloziella turneri** (Hook.) K. Müll.

A species of crumbling slightly acid loam in warm sheltered lightly shaded places, such as rocks on cliffs and beside creeks, stream and ditch banks, and walls and roadside cuttings. It nearly always grows with *Cephalozia bicuspidata*, *Diplophyllum albicans*, *Dicranella heteromalla* and *Polytrichum aloides*, and more rarely with *Epipterygium tozeri*, *Fissidens bryoides* and *F. celticus*. Lowland. GB 21 + 19*, IR 4*.

Autoecious; sporophytes frequent, spring. Gemmae always present.

Mediterranean-Atlantic, reaching Albania, Crete and Scotland. Macaronesia, N. America.

M. F. V. Corley

93

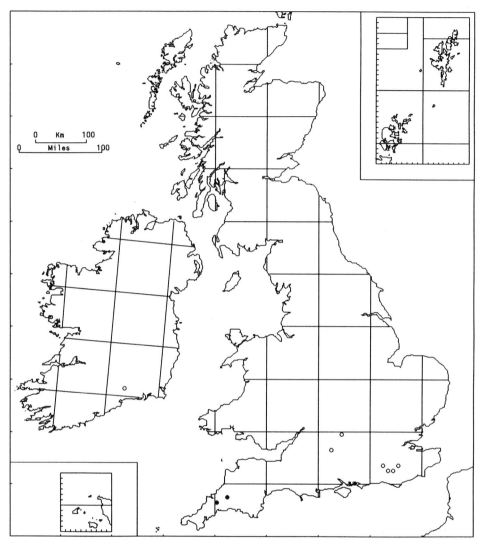

18/12. **Cephaloziella integerrima** (Lindb.) Warnst.

In lowland *Calluna* heath it grows on damp clay soil or sand, especially where compacted by rabbits. In Cornwall it has been recorded from mine waste and from slaty soil in a quarry, and in Sussex and Ireland from roadside banks and a railway bank. Lowland. GB 2+6*, IR 1*.

Autoecious; sporophytes known in Britain. Gemmae common.

In contrast to its British distribution, this is a mainly northern species in Europe, but extends south as far as France and Poland; also recorded from Italy and Hungary.

It appears to be decreasing, probably through loss of lowland heath.

M. F. V. CORLEY

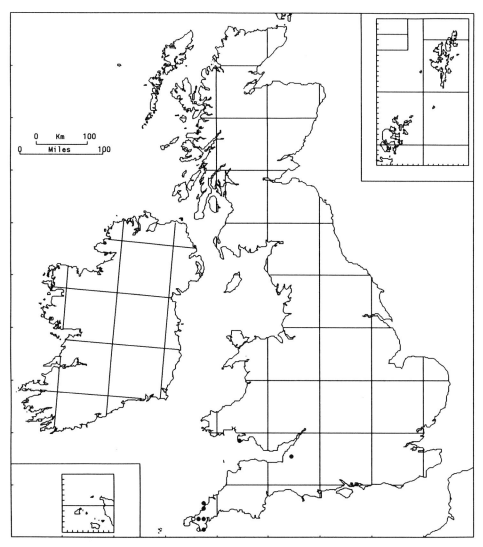

18/13. **Cephaloziella calyculata** (Durieu & Mont.) K. Müll.

A rare plant of thin turf on heathy slopes on sea-cliffs, creek banks and mine waste. Several Cornish sites are on waste from disused copper mines, but those on the Lizard are on serpentine. The Glamorgan and Somerset sites are on limestone. Lowland. GB 7+2*.

Autoecious; perianths regularly produced, sporophytes rare. Gemmae are common.

A western Mediterranean-Atlantic species, extending from the Canaries eastwards to Yugoslavia and northwards to S. Wales.

M. F. V. CORLEY

95

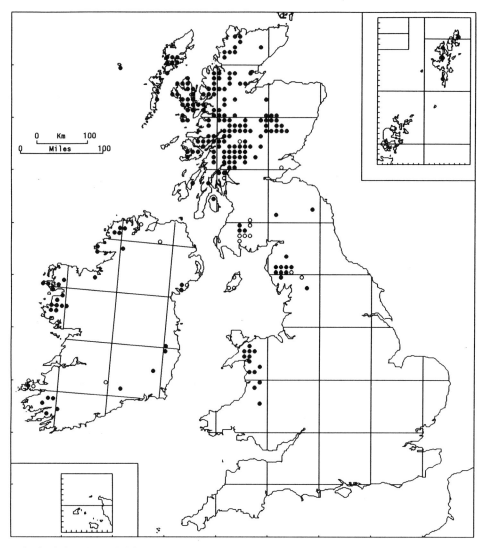

19/1. **Anthelia julacea** (L.) Dum.

A characteristic species of acid flushes, streams and rocks continuously wetted by flowing water, where it often forms swelling masses. In flushed areas in the vicinity of late snow-lie above 850 m it can be the dominant plant (McVean & Ratcliffe, 1962). Most frequent associates are *Marsupella emarginata*, *Scapania undulata*, *Bryum pseudotriquetrum*, *Dicranella palustris*, *Drepanocladus exannulatus* and *Philonotis fontana*. 0–1340 m (Ben Nevis). GB 182+15*, IR 33+6*.

Dioecious; male and female plants frequent but inconspicuous; sporophytes uncommon, maturing during the summer months. Gemmae absent.

Arctic-alpine in Europe from Spitsbergen and N. Scandinavia to the Tatra Mts, Alps and Pyrenees. Also recorded from E. Siberia, W. Himalaya, China, Japan, N. America.

J. G. DUCKETT

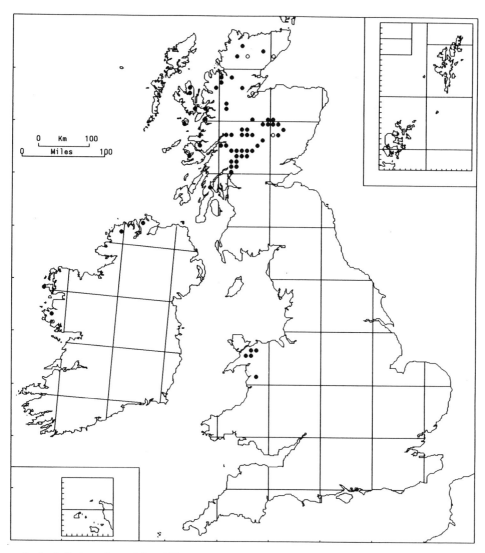

19/2. **Anthelia juratzkana** (Limpr.) Trev.

This species is most abundant at high elevations (above 850 m) in the Scottish Highlands on bare exposed, non-calcareous soils particularly in areas of late snow-lie. Characteristic associates include *Cephalozia* spp., *Gymnocolea inflata*, *Gymnomitrion* spp., *Lophozia* spp., *Marsupella* spp., *Nardia breidleri*, *N. scalaris*, *Dicranella* spp. and *Ditrichum* spp. At lower altitudes it is basicolous on shallow humus-rich soils overlying deeply weathered rocks. Here it grows with *Eremonotus myriocarpus*, *Jungermannia borealis*, *Marsupella boeckii* var. *stableri*, *Anoectangium warburgii* and *Barbula ferruginascens*. 300–1205 m (Ben Lawers). GB 58+5*, IR 4.

Paroecious; fertile plants and sporophytes usually abundant. Gemmae absent.

Arctic-alpine and bipolar: European mountains and throughout the Northern Hemisphere at high latitudes. Also recorded from Bolivia, Argentina, New Zealand, New Guinea, Antarctica.

J. G. DUCKETT

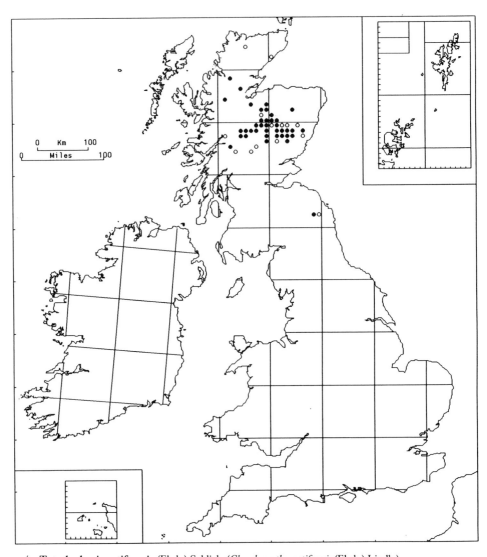

20/1. **Tetralophozia setiformis** (Ehrh.) Schljak. (*Chandonanthus setiformis* (Ehrh.) Lindb.)

It forms conspicuous mats, sometimes in quantity, on dry granite rocks and soil and amongst boulder scree, also rarely on tree roots. 550–1200 m, rarely descending to 275 m. GB 38+14*.

Dioecious; male plants and capsules not known in Britain. Shoots sometimes produce flagelliform microphyllous tips which may act as propagules and the plant may also spread by fragmentation of the brittle stems.

Alpine and Arctic regions of Europe from Spain and Italy north to Spitsbergen. Novaya Zemlya, Siberia, N. America.

A. J. E. SMITH

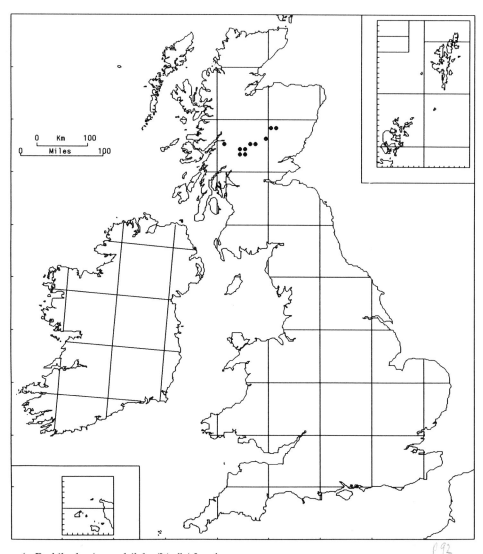

P 92

21/1. **Barbilophozia quadriloba** (Lindb.) Loeske

The very restricted distribution of this species in Scotland reflects its confinement to soft basic schist and metamorphosed limestone at relatively high altitudes. It avoids harder limestone. It usually grows on damp rock ledges, in damp mossy turf or in stony flushes with other calcicoles such as *Saxifraga oppositifolia*, *Selaginella selaginoides*, *Preissia quadrata*, *Bryum pseudotriquetrum*, *Ctenidium molluscum*, *Distichium capillaceum*, *Ditrichum flexicaule*, *Fissidens adianthoides* and sometimes other rare alpine species such as *Scapania degenii* and *Tritomaria polita*. 500–1175 m (Ben Lawers). GB 10.

Dioecious; fertile plants (female) have been found in only one locality. Gemmae, known from Europe, are unrecorded in Britain.

An arctic-alpine, found in Europe in the Alps and Tatra Mts but more abundantly in the tundra of N. Scandinavia, Iceland and Svalbard. Siberia, Arctic America, Greenland.

D. G. LONG

99

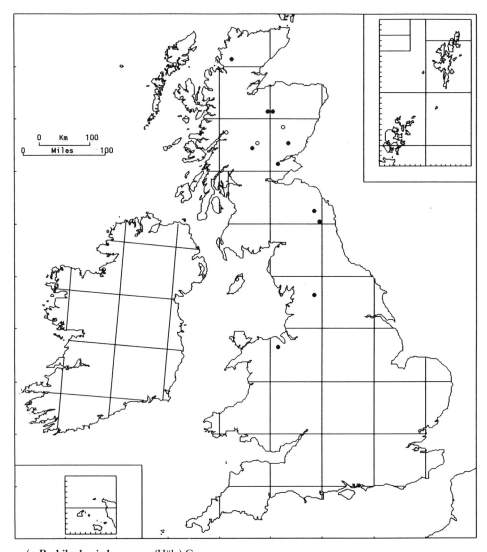

21/2. **Barbilophozia kunzeana** (Hüb.) Gams

Most habitats are damp or waterlogged, and include mountain streamsides, basin mires, wet moorland hollows and valley bogs, associated with *Mylia taylorii, Aulacomnium palustre, Dicranum scoparium, Drepanocladus* spp. and *Sphagnum subnitens*. These associates indicate acidic to mildly basic conditions. At higher elevations on mountains the plant has been recorded from apparently drier rocky slopes and screes. 250–1250 m (Ben Nevis). GB 10+5*, IR 1*.

Dioecious; gametangia very rare, sporophytes unknown in Britain. Gemmae are usually present.

An arctic-alpine, scattered in the European mountains but becoming commoner towards the north, in Scandinavia, Iceland and Svalbard. Siberia, Arctic America, Greenland.

D. G. LONG

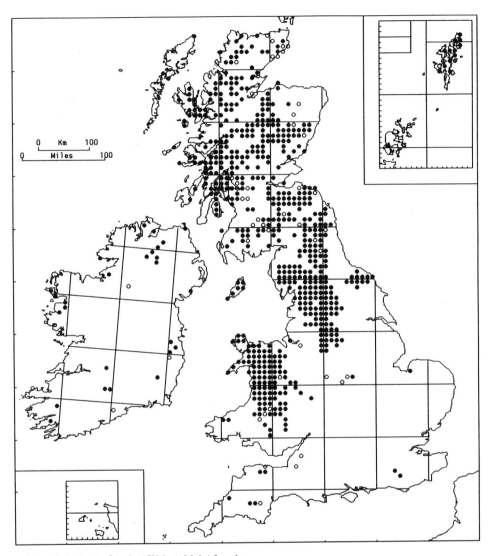

21/3. Barbilophozia floerkei (Web. & Mohr) Loeske

A calcifuge, typical of a variety of well-drained substrates, including fixed coastal dunes, sandy heaths, open moorland slopes under *Calluna*, old quarries, dry screes, rock crevices and drystone walls. The plants often grow as pure stands, or mixed with other calcifuges such as *Dicranum scoparium, Isothecium myosuroides, Pleurozium schreberi* and *Polytrichum* spp. At high altitudes, it is typically found in N.- to E.-facing block-screes and late-snow areas. 0–1220 m (Ben Nevis). GB 585+46*, IR 26+4*.

Dioecious; gametangia uncommon, sporophytes unknown. Gemmae normally absent but occasionally present.

A subarctic-alpine species, widely ranging throughout N. Europe and the Arctic, becoming restricted to mountains in the south. Western and eastern N. America, Peru.

The commonest member of the genus in Britain, but unaccountably rare in Ireland. This species and *B. atlantica* are in need of critical revision; *B. floerkei* may have been over-recorded.

D. G. Long

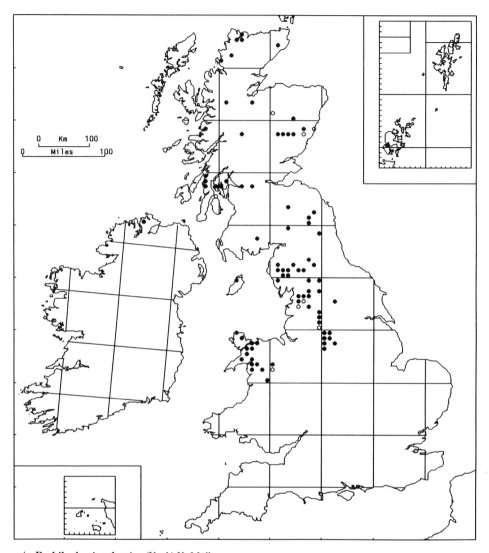

21/4. Barbilophozia atlantica (Kaal.) K. Müll.

Recorded from dry acid block-screes, rocky ravines, lochside boulders and especially from shady sandstone crags and river gorges, with associates such as *Lepidozia cupressina*, *Dicranum scoparium* and *Isothecium myosuroides*. Its habitat is similar to that of *B. floerkei*. Predominantly lowland and subalpine, but to 900 m on Ben Alder. GB 80+8*, IR 1.

Dioecious; gametangia very rare, sporophytes unknown. Gemmae usually present.

A local northern species occurring from France and Switzerland north to Scandinavia, Faeroes and Iceland. Eastern N. America, Greenland.

Possibly under-recorded because of confusion with *B. floerkei*.

D. G. LONG

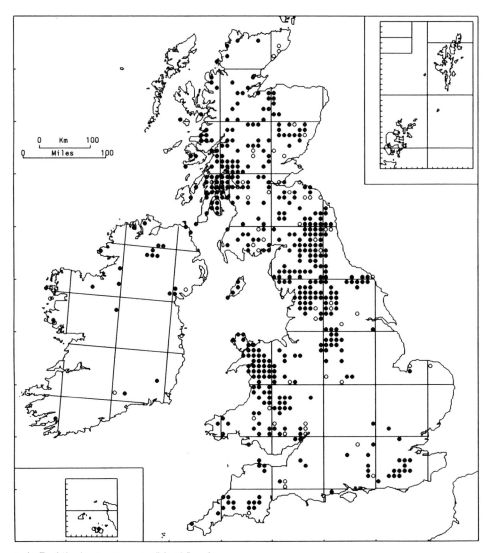

21/5. **Barbilophozia attenuata** (Mart.) Loeske

A lowland and subalpine species characteristic of damp shady woodlands where it grows on logs, stumps, tree-trunks, peaty soil, boulders and walls, often forming extensive pure stands. In the uplands it also grows in block-screes, on sandstone crags and in ravines. Typical associates are *Lepidozia reptans*, *Lophocolea* spp., *Scapania gracilis*, *Dicranum fuscescens*, *D. scoparium* and *Mnium hornum*. 0–550 m (Braemar area). GB 437+54*, IR 23+6*.

Dioecious; male plants frequent, perianths rare, sporophytes unknown in the British Isles. Gemmae invariably present.

North-temperate circumboreal. Widespread in C. and N. Europe, rarer in mountains of the south. Siberia, Japan, Taiwan, N. America, Greenland.

D. G. LONG

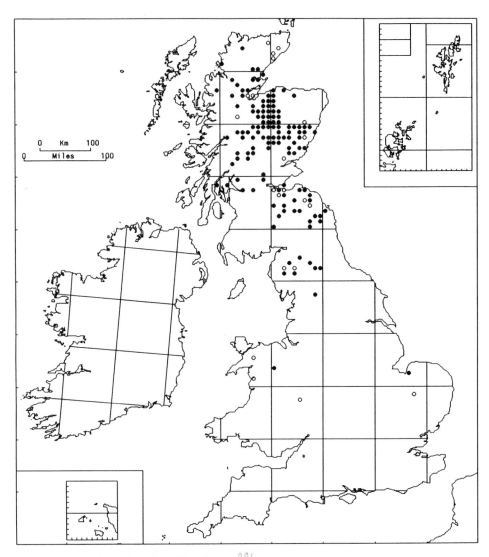

21/6. **Barbilophozia hatcheri** (Evans) Loeske   *P96*

Although more local and eastern than *B. floerkei* in Britain, it has similar ecological requirements. *B. hatcheri* is most commonly found in relatively dry rocky situations, from boulders in woodlands, ravines and open slopes under *Calluna*, to more exposed habitats such as old quarries, dry block-screes and exposed subalpine to alpine cliff ledges. Substrates include acidic rocks (commonly), basic, often igneous rocks such as basalt (more rarely), and open leached mineral soils. Associates are usually common calcifuges such as *Hypnum jutlandicum*, *Pleurozium schreberi*, *Pohlia nutans* etc., but the patches are frequently pure and extensive. Very rarely it grows epiphytically on *Betula*. 75–820 m (Beinn Dorain). GB 127+20*.

Dioecious; male plants common, female plants rare, sporophytes very rare. Gemmae usually abundant.

A circumboreal, arctic-alpine and bipolar taxon. Widespread in Europe north to Svalbard but rare in the south and south-west where it is restricted to mountains. Siberia, Himalaya, N. Asia, Japan, Taiwan, N. America, Greenland, southern S. America, Antarctic Islands.

D. G. LONG

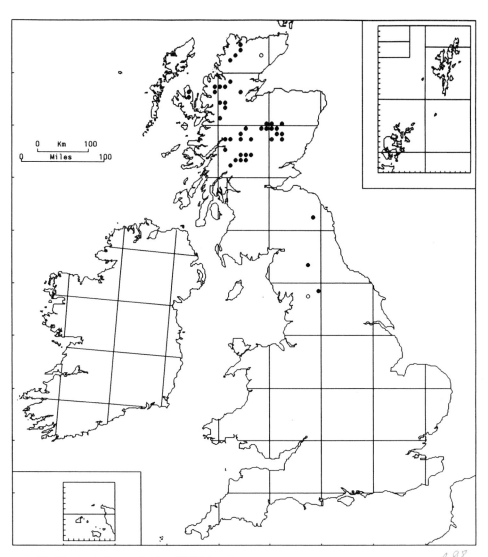

21/7. **Barbilophozia lycopodioides** (Wallr.) Loeske

The distribution of this species in Britain broadly reflects the occurrence of base-rich rocks at moderate to high altitudes, for example on Skye where it is absent from the high-altitude siliceous outcrops but present on the lower basalt hills. It is most common in the Breadalbane mountains on the soft base-rich mica-schists, where it is a frequent component of damp mossy turf on sheltered slopes, on rock ledges and in block-screes. Here, the stems are often scattered amongst a wide range of robust bryophytes such as *Anastrepta orcadensis*, *Tritomaria quinquedentata*, *Dicranum scoparium*, *Drepanocladus uncinatus*, *Hylocomium splendens*, *Hypnum hamulosum* and *Rhytidiadelphus* spp. It can sometimes form almost pure cushions or even robust hummocks, for example on damp rock ledges and in mossy flushes and streamsides, with *Selaginella selaginoides*, *Thalictrum alpinum*, *Bryum pseudotriquetrum* etc. It grows more rarely in base-poor late-snow areas. On Skye it has been reported in oceanic-montane hepatic mats with *Mastigophora woodsii*. 430 m (Skye) to 1175 m (Ben Lawers). GB 44+2*.

Dioecious; gametangia very rare, sporophytes and gemmae unknown in Britain.

Circumboreal arctic-alpine. Widespread on mountains throughout Europe north to Faeroes, Iceland, Scandinavia and Svalbard, and east to the Caucasus. Japan, Kamchatka, N. America, Greenland.

D. G. Long

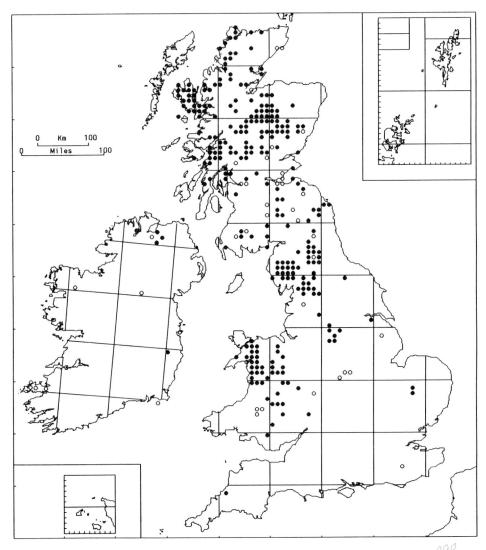

21/8. **Barbilophozia barbata** (Schmid. ex Schreb.) Loeske

Characteristically a lowland to subalpine, calcifuge to mildly calcicolous species growing in dry to moist rocky habitats, particularly on mossy drystone walls, block-screes, rock outcrops, and rock-walls in ravines. Rock types range from acid granite and gneiss to basalt and mildly calcareous sandstone. On rocks and walls it often grows as extensive pure patches, especially where partly shaded. Other habitats include coastal sand-dunes, lowland heaths under Ericaceae and *Ulex* spp., shaded mossy slopes on rocky outcrops, grassy flushes, old pastures, slopes under *Calluna*, and mossy banks in *Betula* and *Pinus sylvestris* woodlands. In such places it grows as scattered stems among other robust bryophytes such as *Barbilophozia floerkei*, *Scapania gracilis*, *Tritomaria quinquedentata*, *Dicranum scoparium*, *Hylocomium splendens*, *Pleurozium schreberi* and *Thuidium tamariscinum*. In woods, *B. barbata* can also grow on stumps and logs, and occasionally epiphytically, e.g. on *Juniperus*. 0–730 m (Loch Avon, Cairngorms). GB 266+33*, IR 4+10*.

Dioecious; male inflorescences and perianths rare, sporophytes not recorded in Britain. Gemmae absent.

Circumboreal. In Europe mainly northern, becoming rare and montane southwards. N. Asia, Japan, N. America, Greenland.

D. G. LONG

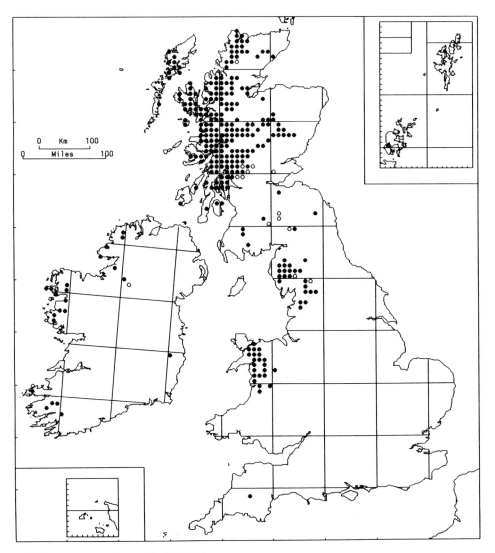

22/1. Anastrepta orcadensis (Hook.) Schiffn.

A distinctive species growing as pure patches or as scattered stems amongst other bryophytes in mixed turf in subalpine *Betula* and *Pinus* woodlands; also frequent on open slopes, mostly in western subalpine to alpine districts, often under *Calluna* or in damp block screes, favouring moist shaded or open N.- to E.-facing aspects. In turfs it is associated with a variety of robust bryophytes such as *Breutelia chrysocoma*, *Dicranum scoparium*, *Plagiothecium undulatum*, *Pleurozium schreberi* and *Rhytidiadelphus loreus*. In rocky woodlands and ravines it often grows with *Hymenophyllum wilsonii* and *Sphagnum quinquefarium*. It is a common component of northern oceanic-montane mixed hepatic mats with *Anastrophyllum donnianum* and similar associates. More rarely it grows amongst sphagnum or on hummocks in blanket bogs. 0–1170 m (Ben Lawers). GB 260+15*, IR 22+3*.

Dioecious; male inflorescences and perianths occasional, sporophytes unknown in Britain. Gemmae common, often abundant in sheltered woodland habitats.

In Europe, mainly in cool oceanic areas of the north and northwest; locally in the high central and eastern mountains. Worldwide it is a disjunct species in cool-temperate high-rainfall areas: Himalaya, W. China, Taiwan, Japan, Aleutians, north-west N. America, Hawaii.

D. G. LONG

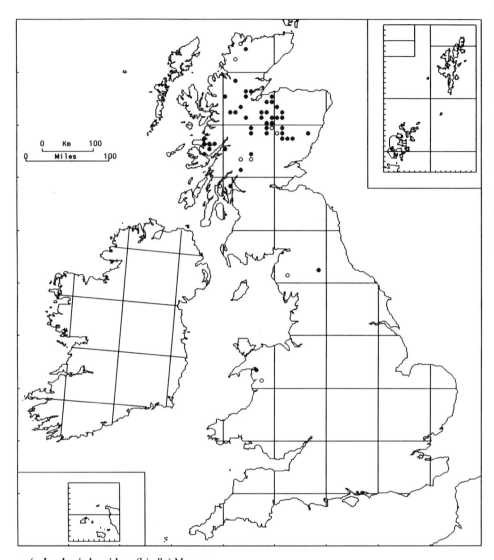

23/1. **Lophozia longidens** (Lindb.) Macoun

Locally common in parts of the eastern Highlands of Scotland, especially in native Caledonian pine forest and mature birch woods, mainly in sheltered situations such as N.-facing slopes, streamsides and ravines. It usually grows on the living bark of trees such as birch and juniper. *Douinia ovata* is a regular associate. It also grows commonly on thin peaty and mossy crusts on the tops and sides of siliceous boulders and rock slabs, both in woodland and on more open but sheltered hillsides; occasionally on more exposed cliff ledges. On rocks and often also on bark, associated species are *Diplophyllum albicans, Frullania tamarisci, Andreaea rupestris, Dicranum scoparium, Hypnum cupressiforme* and *Cladonia* spp. 50 m (Sunart) to 640 m (Beinn Dorain). GB 45+9*.

Dioecious; fertile plants and sporophytes rare. Gemmae usually abundant.

A circumboreal-continental species, found in the mountain ranges of Europe south to N. Italy and Turkey, and north to Scandinavia and Svalbard. Siberia, N. America, China; rare in the Arctic.

D.G. LONG

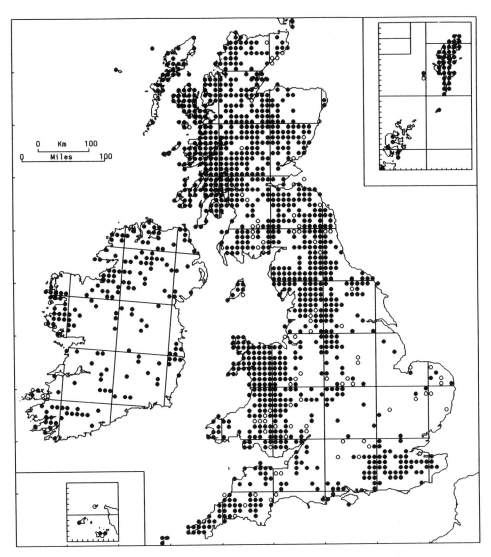

23/2. Lophozia ventricosa (Dicks.) Dum.

A very common species on acid soil or peaty ground. In the north and west it is found in a multiplicity of habitats from coastal sand-dunes, woodlands, mossy walls, logs and stumps, to mine waste, old quarries and open peaty and rocky slopes; in montane districts it is very common in virtually all acidic habitats, especially peaty banks, ditches, sphagnum bogs, dwarf-shrub heaths, screes and cliff ledges. Although a calcifuge, it is frequent on basic rocks where the soil is leached or a peaty or mossy crust has developed. Associates are usually other common species such as *Cephalozia bicuspidata, Gymnocolea inflata, Campylopus* spp., *Ceratodon purpureus, Dicranum scoparium* and *Sphagnum* spp. in more open habitats; and a range of robust bryophytes on woodland banks and sheltered mossy slopes in the uplands. 0–1205 m (Ben Nevis). GB 1135+91*, IR 180+4*.

Dioecious; fertile plants common, sporophytes uncommon. Gemmae usually abundant.

Circumboreal. Common through Europe south to the Mediterranean and Turkey, north to Scandinavia and Iceland but rare in the Arctic. Siberia, N. America, Greenland.

The varieties of this species are too incompletely known for them to be mapped separately.

D. G. LONG

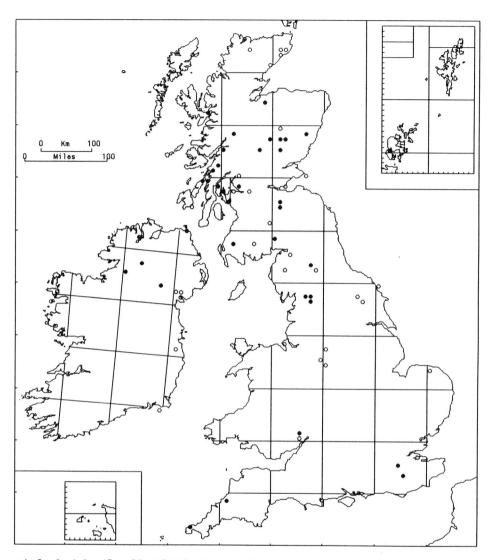

23/3. **Lophozia longiflora** (Nees) Schiffn. (*L. guttulata* (Lindb.) Evans)

The main habitat is on decaying logs and stumps, and rarely peat, in subalpine woods and ravines, sometimes with *Nowellia curvifolia*. Altitudinal range unknown. GB 31+29*, IR 4+6*.

Dioecious; fertile plants and sporophytes reputedly common. Gemmae usually absent.

Pyrenees, Alps, Scandinavia, Svalbard. U.S.S.R., including Siberia and Sakhalin, Korea, N. America, with a C. American disjunction in Guatemala.

A poorly-known species in the British Isles, which was often confused in the past with forms of *L. ventricosa* and which can be reliably distinguished only when bearing perianths. Critical revision of British and Irish material is overdue and the present distribution must be regarded as provisional.

D. G. LONG

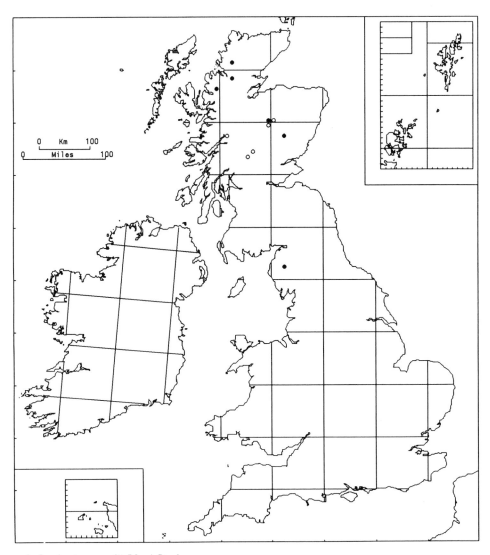

23/4. **Lophozia wenzelii** (Nees) Steph.

Restricted to the higher mountain ranges such as the Cairngorms where it grows in and beside springs, pools and flushes and on wet rocks, especially in late-snow areas. 900–1070 m (Ben Nevis). GB 6+5*.

Dioecious; fertile plants rare. Gemmae usually present.

An arctic-alpine found on the higher mountains of Europe, north to Scandinavia, Faeroes, Svalbard and Iceland. Siberia, Himalaya, Sakhalin, Japan, Taiwan, N. America, Greenland.

It has been confused in the past with forms of *L. ventricosa* and *L. sudetica;* a critical revision of this taxon is needed in Britain.

D. G. LONG

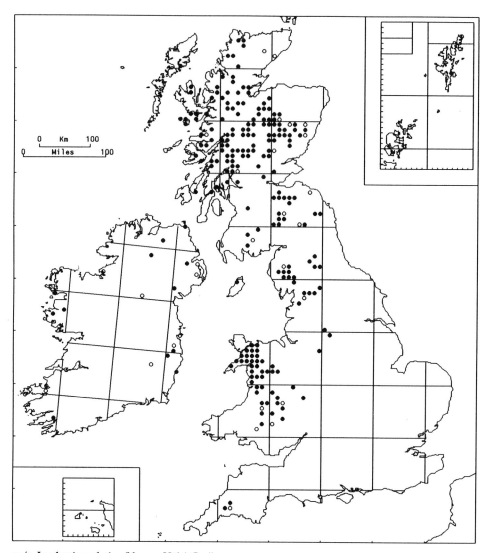

23/5. **Lophozia sudetica** (Nees ex Hüb.) Grolle

Often a common and abundant species on well-drained acidic substrates in exposed montane habitats, e.g. block-screes, rock-faces and boulders, rocky gullies and ravines, footpaths, disturbed ground, late-snow areas, fell-fields, mountain ridges and summits. On rocks it often grows as pure patches; on soil it is often mixed in dwarf mats with other small calcifuge pioneers such as *Diplophyllum albicans*, *Gymnomitrion* spp., *Marsupella* spp., *Nardia breidleri*, *Kiaeria starkei*, *Oligotrichum hercynicum* and *Polytrichum* spp. At lower elevations it is less common but grows generally in dry often rocky situations such as banks by moorland streams, on sun-exposed tops of boulders and walls, in old quarries and on gravelly tracks. On mountains, a mesophytic form is not uncommon in places such as stony flushes and wet rocks and boulders. Substrates are typically acid, but it can grow on more basic rocks such as basalt and sandstone. Sea-level (Moidart) to 1340 m (Ben Nevis). GB 203+22*, IR 12+12*.

Dioecious; gametangia occasional, sporophytes rare. Gemmae often present.

A circumboreal arctic-alpine found in all the higher mountain ranges of Europe, extending south to Turkey, Iberian Peninsula and Madeira. Siberia, Himalaya, Sakhalin, Japan, N. America, Greenland.

D. G. Long

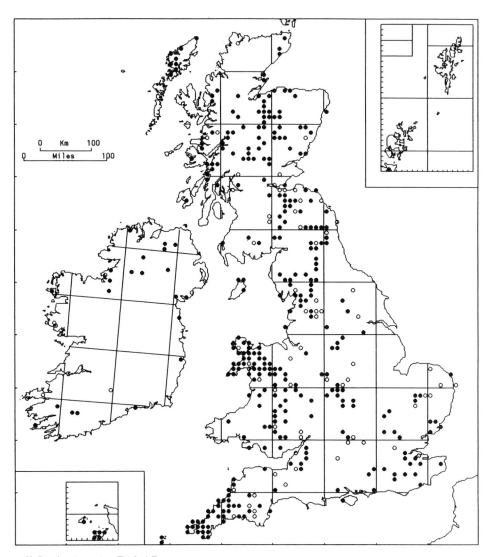

23/6. Lophozia excisa (Dicks.) Dum.

A pioneer species, generally calcifuge, but occasionally found on somewhat basic substrates, or on wall-tops and boulders enriched by bird-droppings. It is most abundant in lowland and subalpine districts. Typical habitats are heathy fixed sand-dunes, rocky banks, roadside quarries, colliery spoil-heaps, peaty wall-tops, sandy and peaty ground on heaths and moorlands, and stony margins of streams, lakes and reservoirs. In the Pennines, it occurs fairly frequently on slightly leached soil on ledges of limestone outcrops. It prefers sun-exposed or partly shaded sites, and is rare in woodland. A range of common calcifuge pioneers are typical associates, for example *Diplophyllum albicans, Jungermannia gracillima, Lophozia ventricosa, Nardia scalaris, Scapania scandica, Ceratodon purpureus, Pohlia annotina, P. nutans, Pogonatum urnigerum, Polytrichum juniperinum* and *P. urnigerum*. 0–600 m (Snowdonia). GB 313+65*, IR 21+6*.

Paroecious; fertile plants and sporophytes common. Gemmae occasional.

Bipolar. In the Northern Hemisphere a circumboreal and arctic species found throughout Europe, north to Scandinavia and the Soviet Arctic. Siberia, N. America, Greenland. In the Southern Hemisphere known from New Zealand, Antarctic Islands and southern S. America.

D. G. LONG

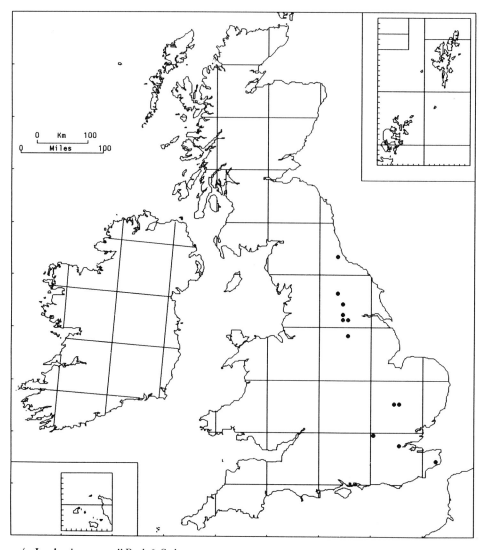

23/7. **Lophozia perssonii** Buch & S. Arn.

A strict calcicole, forming low green patches on disturbed soil and in mossy turf in Magnesian limestone quarries, on bare chalk soil and tracksides in chalk-pits, and in chalk grassland; found once on a limestone wall. Most habitats are relatively exposed and liable to drying out. It has been recorded as growing with the following species: *Aneura pinguis, Leiocolea badensis, L. turbinata, Aloina* spp., *Barbula* spp., *Bryum bicolor, Campylium chrysophyllum, Cratoneuron filicinum, Dicranella varia, Encalypta streptocarpa, Homalothecium lutescens, Seligeria calcarea, Thuidium abietinum, Tortella tortuosa* and *Weissia microstoma*. Lowland. GB 12.

Dioecious; gametangia rare, sporophytes unknown. Gemmae usually abundant.

Endemic to Europe; a rare northern species recorded from France, Denmark, Sweden and Finland.

First discovered in Britain in 1965. Since that date, records have been steadily increasing, and it is likely that the plant is slowly spreading.

D. G. Long

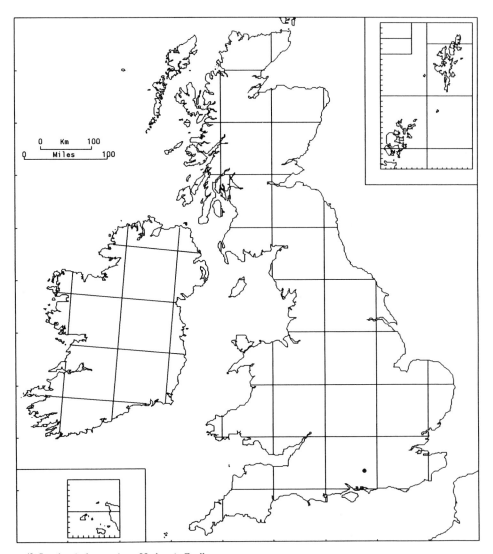

23/8. Lophozia herzogiana Hodgs. & Grolle

Found in 1986 in small quantity growing over *Dicranum scoparium* in a dense Callunetum on Ministry of Defence land. GB 1.

Dioecious. British plants sterile, stunted but abundantly gemmiferous.

Introduced. A native of New Zealand, possibly brought in with troop movements during the 1939–45 war.

A. C. CRUNDWELL

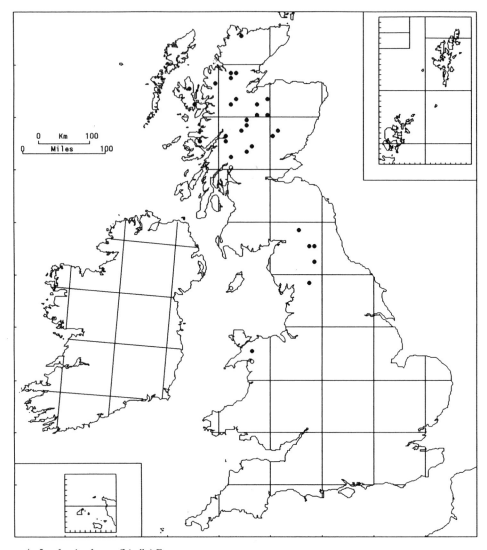

23/9. **Lophozia obtusa** (Lindb.) Evans

Usually growing as scattered stems in mossy turf, this hepatic is often hard to detect in the field and is probably under-recorded. Only rarely, as on river detritus in Northumberland, does it form relatively pure stands. Its rarity is surprising in view of its broad ecological and altitudinal range. At lower elevations it typically grows in woodlands of native pine or birch, scattered amongst bryophytes such as *Diplophyllum albicans, Breutelia chrysocoma, Hylocomium splendens, Pleurozium schreberi, Pseudoscleropodium purum* and *Rhytidiadelphus loreus;* also under *Calluna.* At higher altitudes it can grow with similar associates in mossy, often basic, turf on sheltered N.- or E.-facing slopes. More rarely it grows in basic flushes with *Pellia endiviifolia, Ctenidium molluscum,* etc. On Ben Lawers it grows under mica-schist blocks with *Barbilophozia lycopodioides, Marchantia polymorpha, Bryum pseudotriquetrum* and *Rhizomnium punctatum.* Sea-level (Arrochar) to 1040 m (Ben Lawers). GB 30+3*.

Dioecious; gametangia very rare; sporophytes and gemmae unknown in Britain.

An arctic-alpine, found in most of the mountain ranges of C. and E. Europe, west to Pyrenees and north to Scandinavia, Faeroes and Iceland. Japan, N. America, Greenland.

D. G. LONG

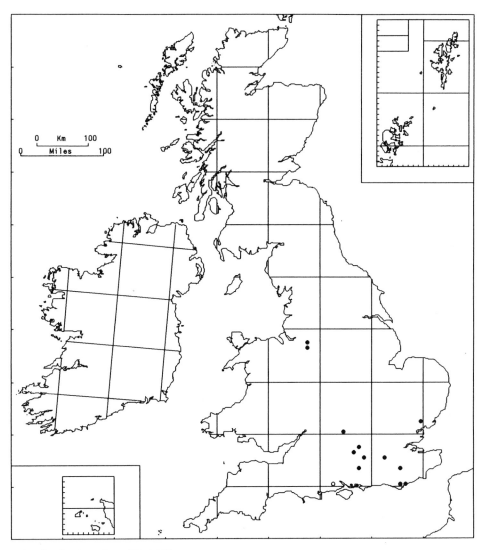

23/10. **Lophozia capitata** (Hook.) Macoun

This lowland calcifuge pioneer grows in green or reddish-tinged patches, or sometimes in quite extensive sheets, on disturbed damp sandy or clay soil, typically on flat ground in old sand-pits where subject to occasional flooding, less often with *Juncus* spp. in open glades or rides in woodland. GB 12+1*.

Dioecious; fertile plants common, sporophytes rare. Gemmae usually present.

Rare in Europe; known from Belgium, Germany, Norway, Sweden and Finland. Eastern N. America.

D. G. Long

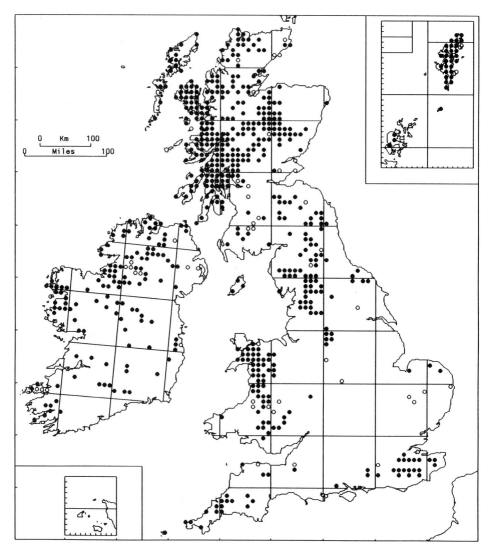

23/11. **Lophozia incisa** (Schrad.) Dum.

A pioneer, occurring in small glaucous-green patches on moist peaty banks, especially in bogs, peat cuttings and damp moorland. It also grows on decaying logs, stumps and litter in damp, open woodland and at the margins of forestry plantations, and on damp humus on river banks and amongst boulders on lake shores. On sandstone and moist shale, it can grow directly on shaded rock-faces. On peat, common associates include *Calypogeia* spp., *Cephalozia* spp., *Diplophyllum albicans*, *Lepidozia reptans*, *Mylia anomala* and *Dicranum scoparium;* on decaying logs additional associates are *Scapania umbrosa*, *Tritomaria exsectiformis* and *Mnium hornum*. 0–640 m (Skye). Many old records from higher altitudes are based on misidentifications of *L. opacifolia*, and the altitudinal limit of *L. incisa* is not known. GB 490+40*, IR 123+13*.

Dioecious; gametangia common, sporophytes rare. Gemmae usually abundant.

A widespread circumboreal species, reported as common in the mountains of Europe (but, as in Britain, some confusion with *L. opacifolia* has probably occurred); frequent in the north and north-west, south to the Pyrenees and Caucasus. Macaronesia, Siberia, Himalaya, E. Asia, N. America, Greenland, C. and S. America.

D. G. LONG

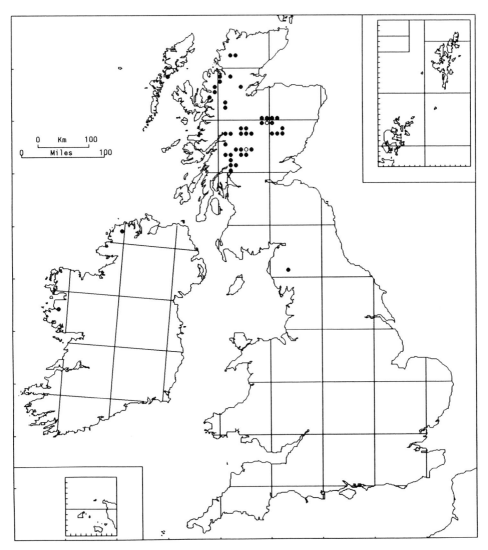

23/12. **Lophozia opacifolia** Culm. ex Meylan

On wet and dripping rocks in gullies and block-screes, especially by waterfalls, springs and flushes, and on irrigated gravel and soil by streams, by lakes and in late-snow areas. It favours cool shaded N.- to E.-facing aspects, and grows with species such as *Anthelia julacea*, *Diplophyllum albicans*, *Marsupella emarginata*, *Moerckia blyttii*, *Nardia scalaris*, *Kiaeria starkei*, *Oligotrichum hercynicum* and *Pohlia nutans*. 400–1140 m (Ben Lawers). GB 42+2*, IR 2.

Dioecious; gametangia and sporophytes occasional. Gemmae usually present.

An arctic-alpine known from the Alps, Scandinavia, Canada and Greenland, but distribution likely to be wider.

This species essentially replaces *L. incisa* at higher altitudes in the mountains of Scotland and Ireland, although it has somewhat different substrate requirements. First detected in Britain in 1956 (Jones, 1957), it is now known to be common on many mountains. Many records of *L. incisa* from above 800 m probably refer to *L. opacifolia*.

D. G. Long

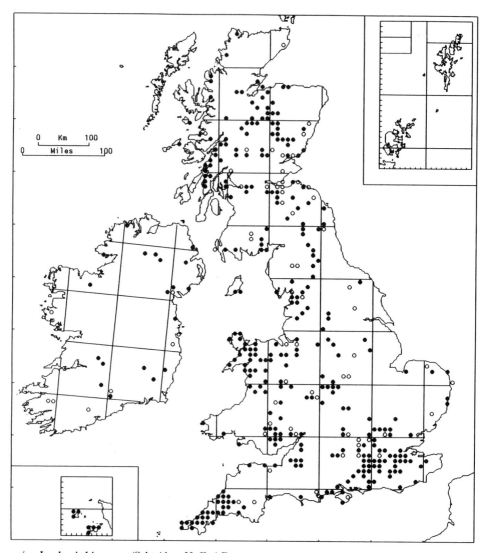

23/13. **Lophozia bicrenata** (Schmid. ex Hoffm.) Dum.

A calcifuge pioneer of open, well-drained peaty and sandy places, such as stable sand-dunes, soil and peat on wall-tops and large boulders, rocky streamsides, compacted footpaths, crumbling banks, old quarries and mine waste. It grows closely attached to its substrate, often under *Calluna*. Associates include a range of pioneer species such as *Diplophyllum albicans*, *Lophozia ventricosa* and *Ditrichum heteromallum*. o–600 m (Snowdonia). GB 304+62*, IR 19+8*.

Paroecious; fertile plants, perianths and sporophytes frequent. Gemmae usually abundant.

In Europe ranging widely from the Mediterranean to the Arctic, but rare in the extreme north. Soviet Asia, N. America.

D. G. Long

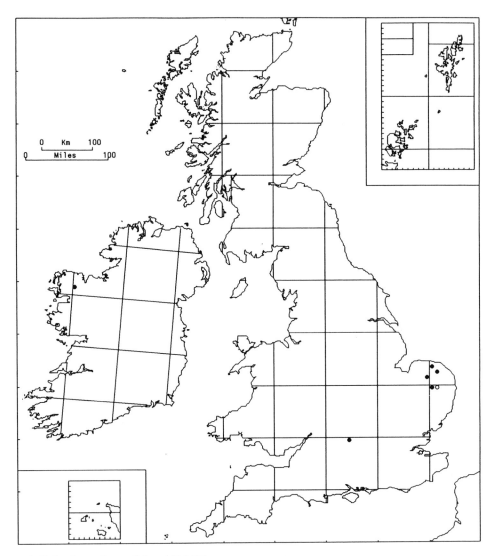

24/1. **Leiocolea rutheana** (Limpr.) K. Müll.

A plant of very wet calcareous lowland fens, with *Carex dioica, Schoenus nigricans, Aneura pinguis, Bryum pseudotriquetrum, Calliergon cuspidatum, Campylium stellatum, Ctenidium molluscum, Drepanocladus revolvens, Homalothecium nitens* and *Scorpidium scorpioides*. In the Irish locality, associated vegetation includes a heathy element with *Myrica gale, Vaccinium oxycoccos* and sphagna. Lowland. GB 5+1*, IR 1.

Paroecious; fertile plants and sporophytes occasional. Gemmae absent.

A circumboreal-continental species generally rare throughout its range, and relictual and threatened in most of Europe. It is recorded from Germany, Denmark, Norway, Sweden, Finland and the U.S.S.R., including Siberia. N. America, Greenland.

L. rutheana has fortunately escaped the fate of *Helodium blandowii, Meesia triquetra* and *Paludella squarrosa*, which are ecologically similar and have become extinct in the British Isles in recent times. Its survival is certainly precarious, and it is probably extinct in some of its E. Anglian localities because of a lowered regional water-table and eutrophication. Its recent discovery in western Ireland (Lockhart, 1989) is therefore especially welcome.

D. G. Long

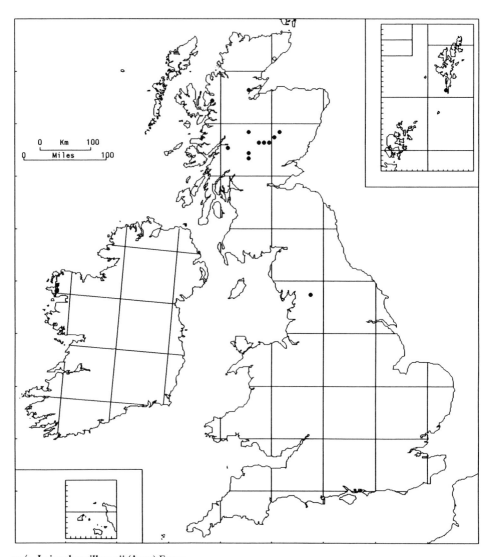

24/2. **Leiocolea gillmanii** (Aust.) Evans

At lower altitudes it is known from damp dune-slacks on Achill Island and in Shetland, and from Old Red Sandstone conglomerate in subalpine ravines in Caithness. In the Breadalbane mountains, it grows on mossy, constantly-irrigated basic rock ledges and in mossy base-rich flushes and rocks by streams. Typical associates are *Selaginella selaginoides, Aneura pinguis, Blepharostoma trichophyllum, Jungermannia* spp., *Preissia quadrata, Bryum pseudotriquetrum, Dichodontium pellucidum, Distichium capillaceum, Fissidens adianthoides, Meesia uliginosa* and *Philonotis* spp. 0–800 m (Breadalbane). GB 13+1*, IR 2.

Paroecious; only known fertile, sporophytes frequent. Gemmae absent.

A circumboreal arctic-alpine taxon found in the mountains of C. and N. Europe. Siberia, N. America, Greenland.

First detected in Britain in Caithness in 1903, and found in Perthshire by H. H. Knight and W. E. Nicholson in 1923. All subsequent localities have been discovered since 1960, other early records having been based on misidentifications of *L. bantriensis*, which is similar in appearance and ecology. It is surprising that *L. gillmanii*, which is more frequently fertile, should be so much rarer than that species.

D. G. LONG

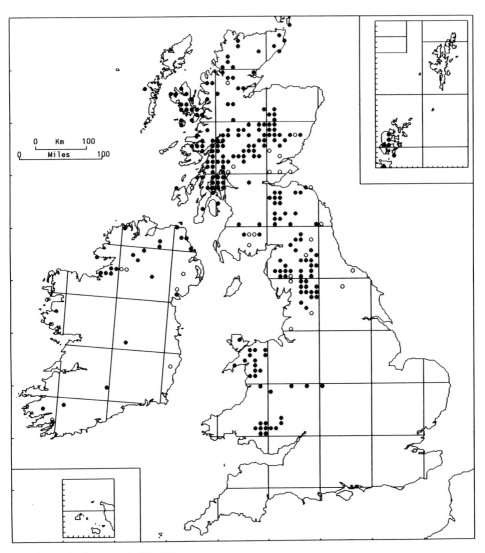

24/3. Leiocolea bantriensis (Hook.) Jørg.

A calcicole which is predominantly montane. At lower elevations it is usually found on constantly irrigated wet rocks in ravines, on rocky streamsides and in lowland or subalpine mires and fens, rarely in coastal dune slacks. In the mountains it grows in characteristic robust tufts in stony flushes, mires and on wet rock ledges, especially on basic schist, limestone and basalt. The patches are often pure, but it commonly grows mixed with other bryophytes such as *Blepharostoma trichophyllum, Pellia endiviifolia, Preissia quadrata, Scapania undulata, Amphidium mougeotii, Blindia acuta, Bryum pseudotriquetrum, Ctenidium molluscum, Distichium capillaceum, Drepanocladus revolvens, Fissidens adianthoides* and *Gymnostomum aeruginosum*. Sea-level (rare) to 820 m. GB 222+23*, IR 23+9*.

Dioecious; fertile plants common, sporophytes rare. Gemmae absent.

A circumboreal arctic-alpine. Widespread in Europe but in the south restricted to mountains, becoming commoner towards the north and west; Scandinavia, Svalbard. Siberia, N. America, Greenland.

D. G. LONG

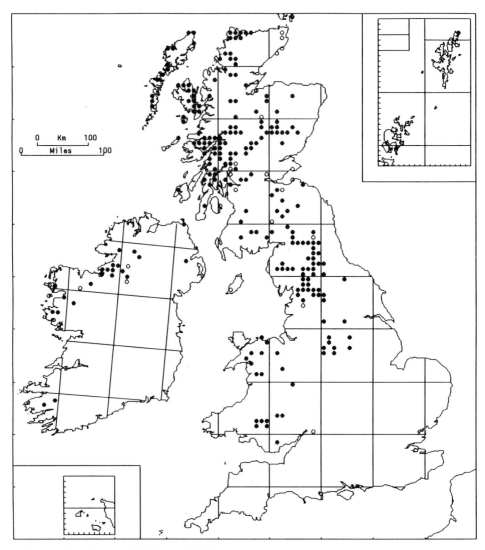

24/4. **Leiocolea alpestris** (Schleich. ex Web.) Isov.

Generally the most frequent member of the genus in base-rich subalpine and montane habitats, although in the north and west it can also grow at lower elevations including coastal sand-dunes. More typically it forms small dark green patches, or grows in mixed bryophyte mats, on soil, rocky flushes (often on boulders), rock ledges in ravines and on mountain cliff ledges, usually where there is a constant seepage of basic water. In northern England, it sometimes grows on shaded boulders in fairly dry woodland. Substrates are often limestone, Old Red Sandstone conglomerate, basic schist and basalt. Plants are commonly mixed with species such as *Jungermannia subelliptica, Preissia quadrata, Bryum pseudotriquetrum, Ctenidium molluscum, Eucladium verticillatum, Fissidens adianthoides, Gymnostomum aeruginosum,* and *Tortella tortuosa.* 0–1175 m (Ben Lawers). GB 210+20*, IR 22+6*.

Dioecious; gametangia frequent, sporophytes rare. Gemmae absent.

Distributed throughout Europe, especially in calcareous areas in the mountains, north to Scandinavia, Faeroes and Iceland. Throughout the Arctic; N. America, Greenland.

D. G. LONG

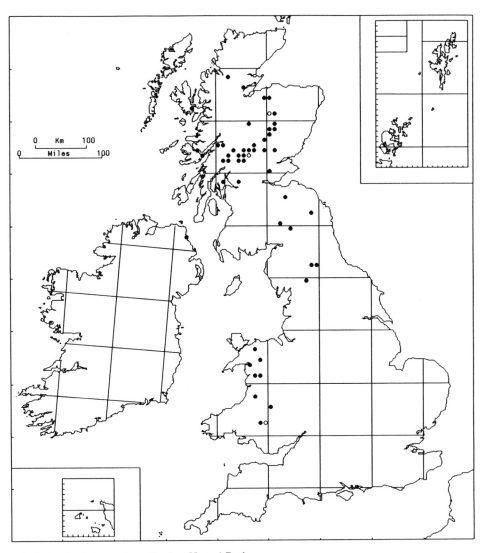

24/5. Leiocolea heterocolpos (Thed. ex Hartm.) Buch

A subalpine and alpine hepatic growing usually on damp calcareous rock outcrops in ravines, on streamsides and on shaded cliffs, typically creeping over cushions of other bryophytes, especially *Blepharostoma trichophyllum*, *Scapania aequiloba*, *Amphidium mougeotii*, *Anoectangium aestivum*, *Distichium capillaceum*, *Gymnostomum aeruginosum*, *Orthothecium rufescens* and *Trichostomum crispulum*, often as scattered stems in small quantity. 100 m (Ceunant Llennyrch) to 900 m (Cruach Ardrain). GB 49+3*, IR 1.

Dioecious; gametangia and sporophytes rare. Gemmae frequent.

Mountain ranges throughout Europe south to Pyrenees and Madeira, but commoner towards the north, especially Scandinavia; also in Svalbard and Iceland. U.S.S.R., Himalaya, N. America, Greenland.

About half the records have been added in the past two decades, and it is probably still under-recorded.

D. G. LONG

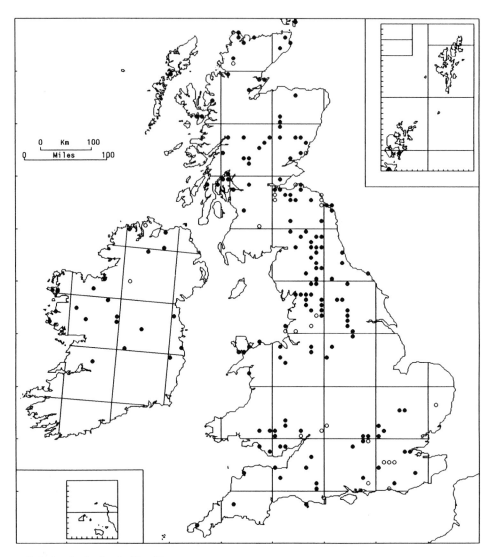

24/6. **Leiocolea badensis** (Gott.) Jørg.

Similar ecologically to *L. turbinata* but more widely ranging, from coastal dune-slacks to damp spots on calcareous rocks in woodlands, rocks by streams, damp basic soil, and basic rocks in subalpine flushes. On the Magnesian limestone in Yorkshire it is characteristic of quarry spoil. It is much less common than *L. turbinata* in chalk and limestone grassland. 0–370 m (Glen Lochay). GB 151+24*, IR 21+3*.

Dioecious; gametangia and sporophytes common. Gemmae absent.

A boreal-alpine species found throughout Europe but rare in the south where it is restricted to mountains. N. Asia, Iran, N. America, Greenland.

D. G. Long

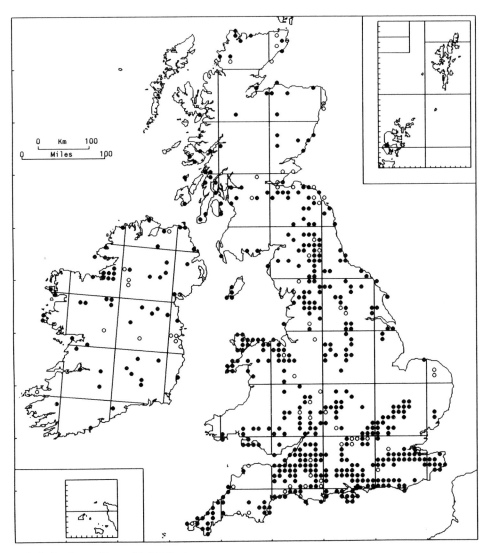

24/7. Leiocolea turbinata (Raddi) Buch

A strict calcicole, growing as thin green patches in moist dune-slacks, and on moist basic sea banks and cliffs. Inland it is common on limestones, chalk and Carboniferous sandstones, often in seepages or on shaded humid rocks and boulders in woodland. On these rock types it is also common on rocky river banks, in ravines, on walls (usually on damp mortar), in limestone quarries and chalk pits (on both rock and soil) and on soil in limestone and chalk grassland. It grows with a range of other calcicolous bryophytes such as *Aneura pinguis*, *Jungermannia atrovirens*, *Pellia endiviifolia*, *Barbula tophacea* and *Dicranella varia*. 0–380 m (Llangattock). GB 456+57*, IR 56+12*.

Dioecious; male inflorescences and perianths common, sporophytes occasional. Gemmae absent.

A Mediterranean-Atlantic species, common around the Mediterranean, including N. Africa, and through much of W. Europe to W. Germany and Belgium. In Asia extending to Iran. Disjunct in Guatemala.

D. G. Long

127

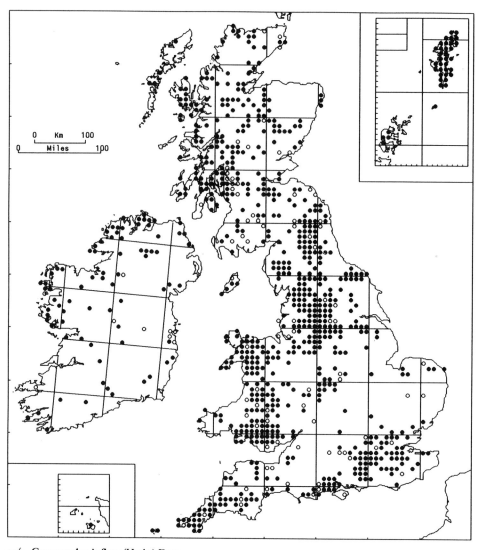

25/1. **Gymnocolea inflata** (Huds.) Dum.

On a wide range of substrata, including peat, rock, sand, and clay, always where strongly acid and at least moderately well illuminated, often where few other plants except *Calluna*, *Campylopus* spp. and *Pohlia nutans* can survive. Its most characteristic habitat is wet heath, where it is sometimes abundant on thin peat at the edge of temporary pools, with *Sphagnum* spp. such as *S. compactum* and *S. tenellum*, as well as the associates mentioned above; but it also grows on eroding blanket peat, rocky banks and intermittently irrigated rocks. Very tolerant of pollution, it often occurs near large cities and on toxic waste from copper mines and other extractive industries. 0–1100 m (Cairngorms). GB 704+83*, IR 77+9*.

Dioecious. Female plants are common and dispersed by caducous perianths. Male plants are occasional; they lack a specialized means of dispersal. Sporophytes are very rare.

Throughout N. and C. Europe. Widespread elsewhere, including Azores, N. America and E. Asia.

M. O. HILL

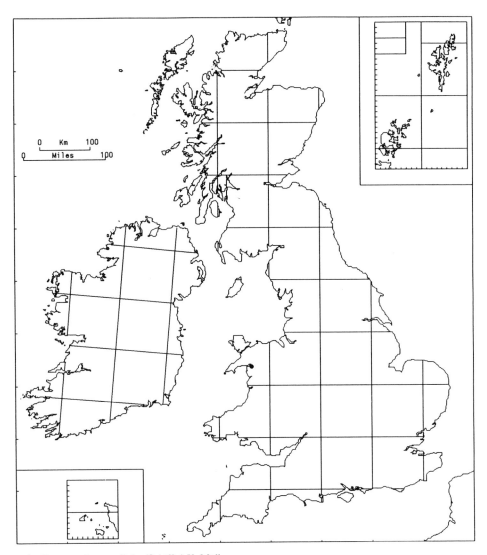

25/2. Gymnocolea acutiloba (Schiffn.) K. Müll.

On boulders and rocks in block scree; confined to a part of the Rhinog mountains noted for its rich bryophyte flora. The rocks are hard and crystalline and the habitat resembles that of montane occurrences of *G. inflata*. The main associates are lichens, notably *Cladonia arbuscula, C. portentosa* and *Sphaerophorus fragilis*, with the liverwort *Anastrepta orcadensis* in crevices. 600 m. GB 1.

Dioecious; female in Britain, but perianths are rarely produced and do not provide an effective means of dispersal.

N. Europe and mountains of C. Europe, rare throughout its range. Outside Europe only in Maine (U.S.A.) and Greenland.

M. O. HILL

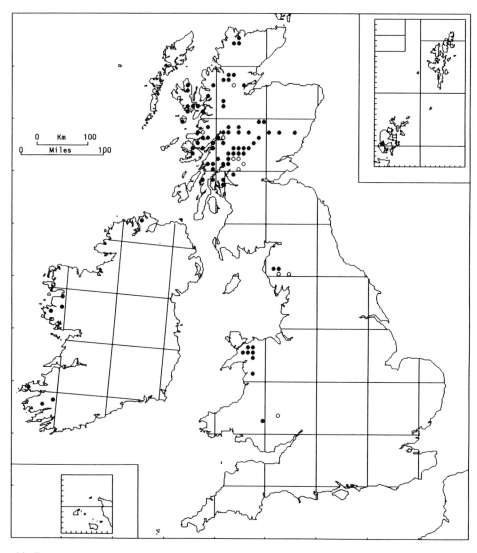

26/1. **Eremonotus myriocarpus** (Carringt.) Lindb. & Kaal. ex Pears.

It usually occurs as pure patches on periodically flushed schistose or other basic rocks and on small damp ledges of shaded cliffs, growing over thin, soft lithosol or mixed with detritus. It favours N. or E. aspects. It also occurs on basic rocks in ravines, particularly on the shaded vertical walls of deep ravines and on rocks beside streams and rivers. At higher elevations it has been reported growing as cushions on the surface of more exposed rocks (Paton, 1966). It occurs from near sea-level to 1200 m, but is most frequent above 450 m. GB 75+11*, IR 7.

Dioecious; perianths uncommon, capsules rare.

Widespread but local in subalpine and alpine areas in Europe, extending from Bulgaria and Italy northwards to Svalbard. E. Asia, Greenland.

H. J. B. Birks

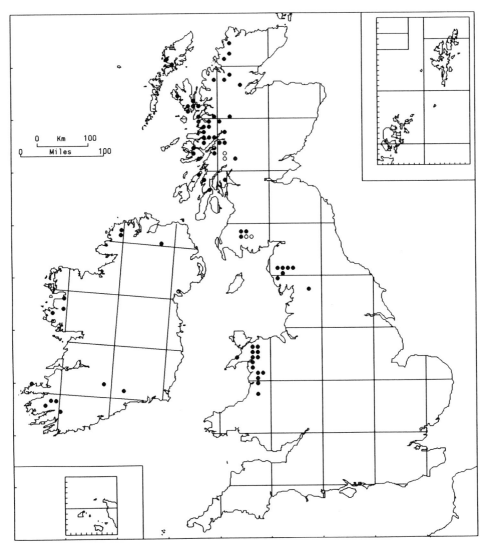

27/1. Sphenolobopsis pearsonii (Spruce) Schust.

It grows in minute appressed green to blackish mats or as scattered stems on damp shaded rock-walls and boulders in sheltered ravines, on large blocks in scree, and on vertical rock-walls on montane cliffs with a N. to E. aspect. Rocks are usually acid, e.g. granite and gneiss. The plants may grow directly on the rock or on a thin peaty or mossy crust, sometimes mixed with filamentous algae. Associates include *Diplophyllum albicans* (commonly), *Jungermannia* spp. and *Marsupella emarginata*. 0–890 m (Ben Wyvis). GB 64+5*, IR 13+2*.

Dioecious; male plants frequent, female plants rare but scattered through Britain and Ireland, sporophytes not found. Gemmae absent.

A highly disjunct species of cool-temperate high-rainfall areas: Norway; Nepal, Bhutan, Taiwan, British Columbia, Appalachian Mts.

The distribution of this species both in the British Isles and globally shows many similarities with that of the larger disjunct oceanic-montane liverworts, but it is ecologically quite different.

D. G. LONG

131

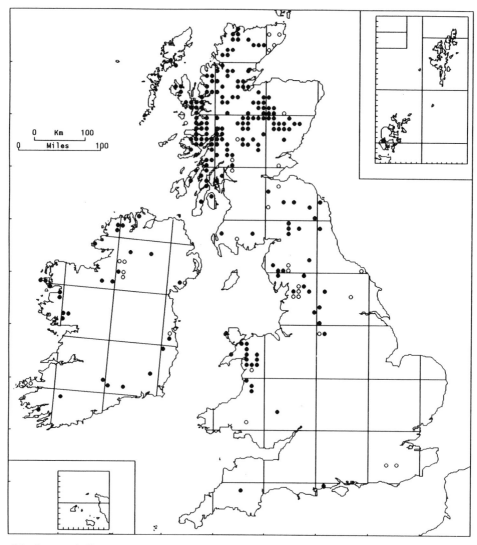

28/1. **Anastrophyllum minutum** (Schreb.) Schust. (*Sphenolobus minutus* (Schreb.) Berggr.)

A calcifuge growing in a wide variety of sheltered habitats: peaty banks; shaded acid rock faces; tree stumps and sometimes living bark in woodland and ravines; and acid bogs, where it can grow over sphagnum and on drier hummocks. In subalpine and alpine sites, it grows in more exposed situations such as cliff ledges, boulders and open peaty banks, often in mixed bryophyte mats under *Calluna*. A broad spectrum of rock types is suitable; sandstones are especially favoured. On basic substrates, even limestone, it will grow where a thin peaty crust has developed. 0–1000 m (Ben Lawers). GB 190+29*, IR 27+9*.

Dioecious; male inflorescences and perianths common, sporophytes rare. Gemmae occasional.

Circumboreal and tropical-alpine. Mountains of Europe south to Azores, Pyrenees and Caucasus; commoner and at lower altitudes in the north and west. Siberia, Himalaya, N. America, Greenland, tropical African mountains, S. Africa, Borneo, Papua New Guinea, Mexico, Venezuela.

Its patchy distribution is not easily explained; it is quite rare in some apparently suitable districts.

D. G. LONG

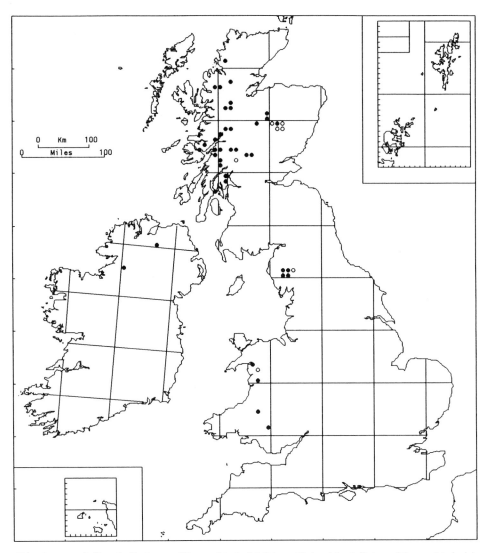

28/2. **Anastrophyllum hellerianum** (Nees ex Lindenb.) Schust. (*Sphenolobus hellerianus* (Nees ex Lindenb.) Steph.)

This dwarf and often inconspicuous species inhabits damp rotting, usually decorticated logs and stumps in shady situations in subalpine woodlands and wooded ravines. It is especially characteristic of native Caledonian pine-forests, but is rarely found in conifer plantations. It also grows on living bark of *Quercus* and on stems and twigs of *Juniperus*. In Argyll it has been found growing in crevices of large scree-blocks where a thin peaty crust has developed. Common associates are *Calypogeia* spp., *Cephalozia* spp., especially *C. catenulata* and *C. lunulifolia*, *Nowellia curvifolia*, *Riccardia palmata*, *Scapania umbrosa* and *Cladonia* spp. 45–250 m (Glen Affric). GB 37+7*, IR 2.

Dioecious; gametangia rare, sporophytes unknown in Britain. Gemmae usually present and abundant.

A circumboreal-subcontinental species, widespread in the mountainous parts of C. and N. Europe and eastern N. America, with scattered occurrences in Siberia, Bhutan, Japan, western N. America and Mexico.

D. G. LONG

133

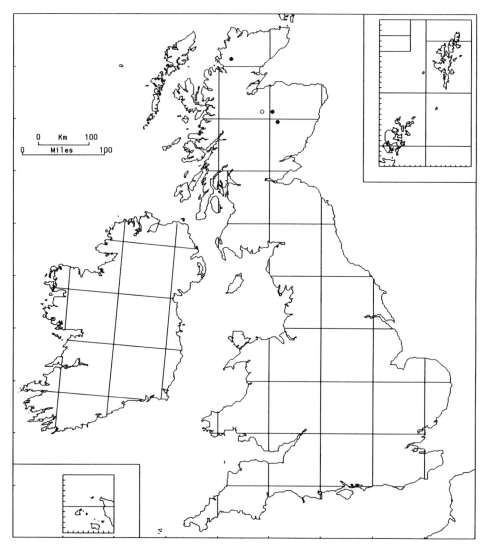

28/3. **Anastrophyllum saxicola** (Schrad.) Schust.

In its best-known locality it grows on dry, N.-facing, acid scree on a slope with scattered native *Pinus sylvestris* trees above a small loch. Other records also suggest scree or boulder habitats. Associated species are *Tetralophozia setiformis*, *Cynodontium jenneri*, *C. strumiferum*, *Racomitrium lanuginosum* and a range of lichens including *Cladonia* spp., *Coelocaulon aculeatum*, *Hypogymnia physodes*, *Sphaerophorus globosus* and *Stereocaulon vesuvianum*. 350 m (Cairngorms). GB 3 + 1*.

Dioecious; gametangia not found in Britain. Gemmae absent.

A circumpolar arctic-alpine with a distinctly continental range in Europe; uncommon in the mountains of C. and E. Europe, becoming more frequent in Scandinavia. Siberia, Japan, N. America, Greenland.

D. G. LONG

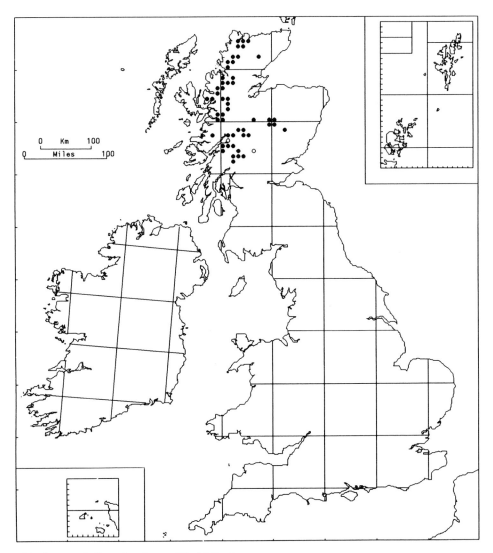

28/4. **Anastrophyllum donnianum** (Hook.) Steph.

A distinctive hepatic of well-drained slopes, especially where shade is continuous for much of the winter and snow lies late. It often grows in N.- and E.-facing mountain corries, either where large loose blocks have accumulated, creating locally sheltered, humid conditions, or on rocky slopes close to water. It also occurs on more open slopes under *Calluna* and other dwarf shrubs, and on mossy ledges of acid cliffs where it may be regularly irrigated by run-off from rocks. It has a wide range of associates, including other conspicuous oceanic-montane hepatics such as *A. joergensenii*, *Bazzania pearsonii*, *Herbertus aduncus*, *Mastigophora woodsii*, *Pleurozia purpurea* and *Scapania ornithopodioides*, but the mats are frequently co-dominated by commoner species, e.g. *Anastrepta orcadensis*, *Bazzania tricrenata*, *Diplophyllum albicans*, *Mylia taylorii*, *Pleurozium schreberi*, *Racomitrium lanuginosum* and *Rhytidiadelphus loreus*. 500 m (Beinn Eighe) to 1060 m (Ben Nevis). GB 51+2*.

   Dioecious; perianths and male inflorescences occasional; sporophytes found only once. Gemmae unknown.

   A disjunct species of temperate high-rainfall areas: Faeroes, S.W. Norway; W. Tibet, Sikkim, Nepal, Bhutan, Yunnan, Alaska, W. Canada.

D. G. Long

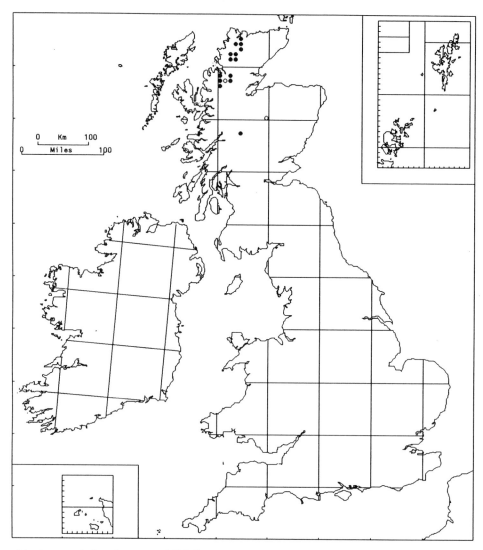

28/5. **Anastrophyllum joergensenii** Schiffn.

It is abundant only on a few hills in Wester Ross and Sutherland, where it occupies similar habitats to, and often occurs intermixed with, *A. donnianum*. It grows in relatively exposed boulder-fields on Ben More Assynt, and has been reported from shallow high-level blanket bog. In the Ben Alder range, the habitat is a rocky outcrop close to a corrie loch. Associates are similar to those of *A. donnianum*. 730–950 m (Beinn Dearg). GB 14+2*.

Dioecious; gametangia, sporophytes and gemmae unknown.

A relict disjunct species of cool montane high-rainfall areas: S.W. Norway; Nepal, Sikkim, Bhutan, Yunnan.

Like other disjunct oceanic-montane hepatics (e.g. *Mastigophora woodsii, Plagiochila carringtonii, Pleurozia purpurea* and *Scapania ornithopodioides*) in the eastern Himalaya it is characteristic of moist juniper-rhododendron scrub just above the tree-line, and on mossy rocks, logs and tree-trunks in montane fir-rhododendron forests. It is possible that in Scotland these hepatics formerly grew at or near the natural tree-line and in associated dwarf-shrub heaths. This transition is now almost totally lost in Scotland through a combination of clearance, burning and grazing.

D. G. LONG

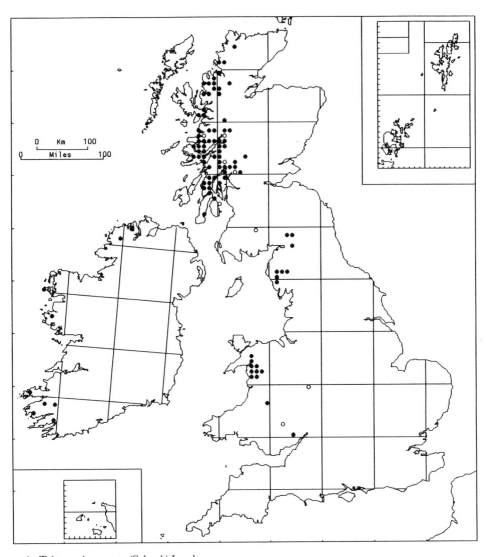

29/1. **Tritomaria exsecta** (Schrad.) Loeske

A lowland plant restricted to woodlands, especially in sheltered ravines, where it inhabits stumps and logs, more rarely mossy rocks and peaty banks, with similar associates to *T. exsectiformis*. Unlike that species it usually grows in small quantity. o–38o m (Snowdonia). GB 84+8*, IR 9+2*.

Dioecious; gametangia very rare, sporophytes unknown in Britain. Gemmae invariably present.

Circumboreal and disjunct tropical-alpine. In continental Europe a subalpine and alpine species: Pyrenees; widespread in the central and northern mountains to S. Scandinavia, but absent from the north and the Arctic. Madeira; scattered through Asia to Japan, Taiwan, Philippines and Borneo; widespread in N. America south to the southern Appalachians; disjunct in Mexico and the E. African mountains.

D. G. Long

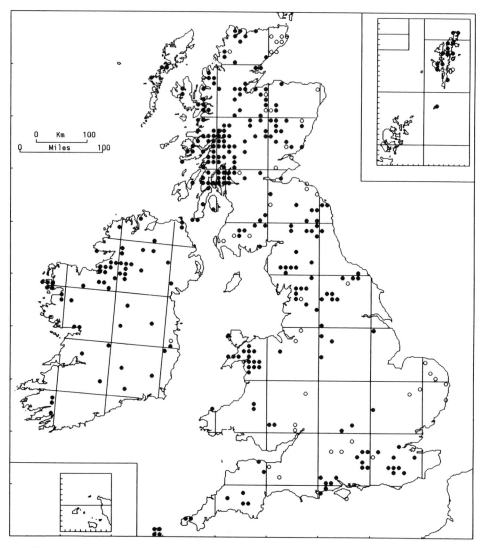

29/2. **Tritomaria exsectiformis** (Breidl.) Loeske

A lowland and subalpine species most often growing in damp woodland, both coniferous and broadleaved. It is frequent on decaying decorticated logs and stumps, where it often accompanies *Calypogeia* spp., *Diplophyllum albicans*, *Lophozia ventricosa*, *Nowellia curvifolia*, *Riccardia palmata*, *Scapania gracilis*, *S. umbrosa*, *Campylopus pyriformis* and *Hypnum cupressiforme*. It is sometimes abundant in other habitats, notably on peaty crusts on sandstone rocks, on boulders in rocky woods and on peaty wall-tops. It occasionally grows on bare peat on damp hillsides and in bogs and peat cuttings. 0–500 m (Glen Clova). GB 255+50*, IR 56+1*.

Dioecious; male inflorescences and perianths rare, sporophytes not recorded in the British Isles. Gemmae abundant and invariably present.

A circumboreal species. Widespread through Europe, montane in the south, becoming rare in the far north; Arctic populations probably belong to a distinct subspecies. Caucasus, Siberia; widespread in N. America; S. Greenland.

D. G. Long

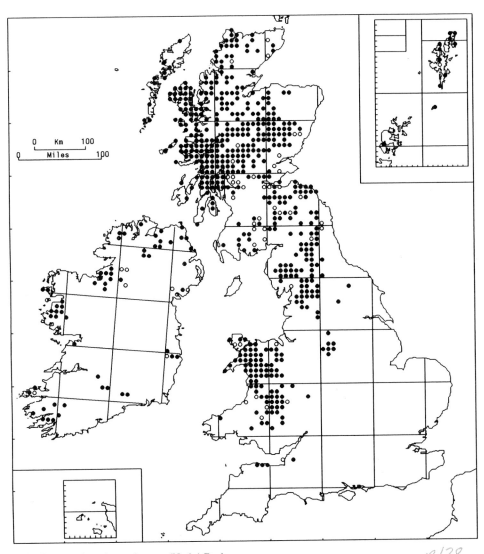

29/3. **Tritomaria quinquedentata** (Huds.) Buch

Common in a wide diversity of habitats including lowland and subalpine woods, bogs, mountains and, rarely, sand-dunes. In woodland it can grow on mossy banks, limestone boulders, shaded walls, tree-trunks, stumps and rocky streamsides. At higher altitudes it occurs in mixed bryophyte mats under *Calluna*, and on rock outcrops, block-screes and cliff ledges. In turf, it grows as scattered stems amongst other large bryophytes such as *Diplophyllum albicans*, *Plagiochila spinulosa*, *Dicranum scoparium*, *Hylocomium splendens* and *Mnium hornum*. In the mountains it appears to be more abundant on basic substrates such as cliff ledges with *Blepharostoma trichophyllum*, *Anoectangium aestivum*, *Amphidium mougeotii* etc. In bogs it can grow mixed with sphagnum. 0–1205 m (Ben Lawers). GB 510+45*, IR 59+9*.

Dioecious; gametangia frequent, sporophytes occasional. Gemmae sporadic.

Circumboreal arctic-alpine. Widespread in Europe, especially in the mountains, north to Faeroes, Iceland and Svalbard. Soviet Asia, N.W. Himalaya, China, Taiwan, Japan, N. America, Greenland.

D. G. LONG

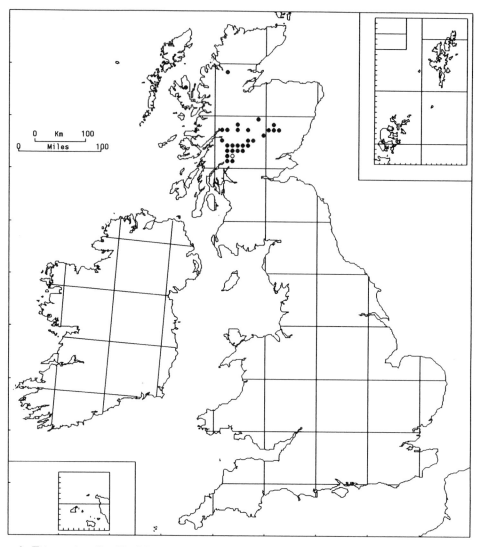

29/4. Tritomaria polita (Nees) Jørg.

Despite its restricted range in Britain this strongly calcicolous species is locally abundant, particularly in the Breadalbane mountains. It forms characteristic greenish often pure patches on streamsides, in stony flushes and on wet gravelly cliff-ledges, favouring N.- or N.E.-facing slopes. The substratum may be mica-schist, metamorphosed limestone or basalt. Common associates include *Juncus triglumis*, *Selaginella selaginoides*, *Thalictrum alpinum*, *Leiocolea bantriensis*, *Bryum pseudotriquetrum*, *Calliergon sarmentosum*, *Ctenidium molluscum*, *Drepanocladus revolvens*, *Fissidens adianthoides* and *Philonotis fontana;* rarer associates are *Carex atrofusca*, *Harpanthus flotovianus*, *Jungermannia borealis* and *Bryoerythrophyllum caledonicum*. 400 m (Glen Coe) to 1070 m (Breadalbane). GB 28+1*.

Dioecious; gametangia occasional, sporophytes rare. Gemmae not recorded in Britain.

An arctic-alpine found in the mountains of C. Europe north to Scandinavia, Svalbard, U.S.S.R.; N. America, Greenland.

D. G. Long

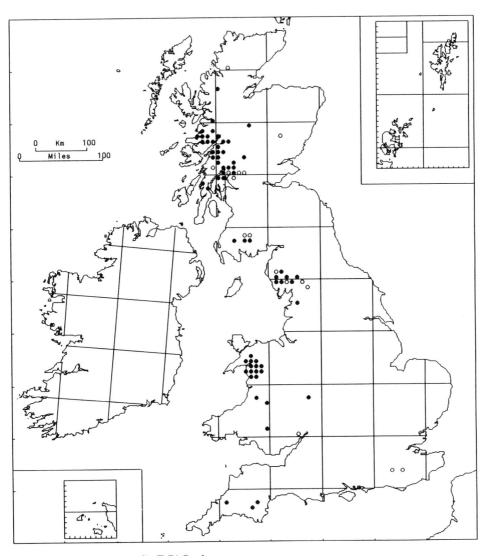

30/1. **Jamesoniella autumnalis** (DC.) Steph.

This species favours damp, decaying decorticated logs and stumps in oak or birch woods on steep, block-strewn slopes with a range of aspects. Associates commonly include *Cephalozia* spp., *Nowellia curvifolia*, *Riccardia palmata*, *Tritomaria exsecta* and *T. exsectiformis*, and, more rarely, *Anastrophyllum hellerianum*. It also occurs on damp, shaded acid or mildly basic blocks and low rock outcrops that are periodically flushed in sheltered oak or mixed oak-birch woods, wooded valleys and on steep slopes, growing with *Harpanthus scutatus*, *Scapania umbrosa*, *Tritomaria exsecta* and *Sematophyllum* spp. Restricted to low elevations (0–350 m). GB 62+17*.

Dioecious; perianths frequent, capsules rare.

Occurs widely but locally in Fennoscandia, W. and C. Europe, Caucasus, Siberia, Japan, N. America.

Its absence from Ireland is curious. There are several old records that have not been confirmed recently, particularly in the Weald and S.W. Scotland, possibly due to changes in woodland management and conifer planting.

H. J. B. BIRKS

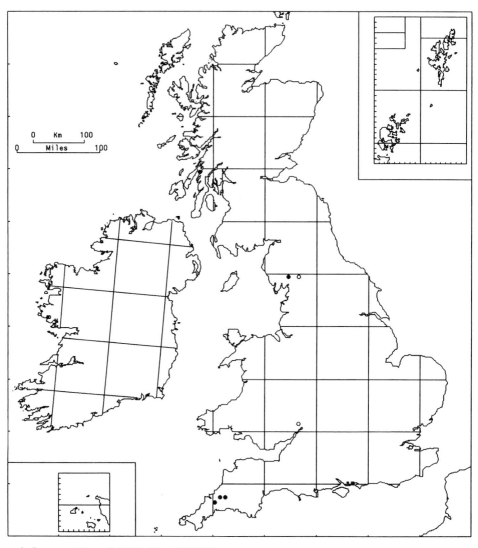

30/2. **Jamesoniella undulifolia** (Nees) K. Müll.

In its English localities it occurs in small quantity in wet minerotrophic sphagnum mires, growing with *Juncus acutiflorus, J. effusus, Sphagnum recurvum, S. teres, S. squarrosum* and *S. subnitens*. In western Scotland it has been found in a large grassy sphagnum tussock, mixed with *Dicranum scoparium* and *Sphagnum capillifolium*. It occurs at low elevations but extends to nearly 300 m in Cornwall. GB 5+3*.

Dioecious; perianths rare.

A suboceanic species occurring widely but locally throughout Europe, from Czechoslovakia to Fennoscandia. It is also known from E. Asia and N. America.

It is curiously rare in Britain in view of the widespread occurrence of its habitat. Such habitats are increasingly disappearing because of drainage and agricultural improvement. In one of its sites on Bodmin Moor, it was locally abundant until a few years ago when it was drowned by a reservoir.

H. J. B. Birks

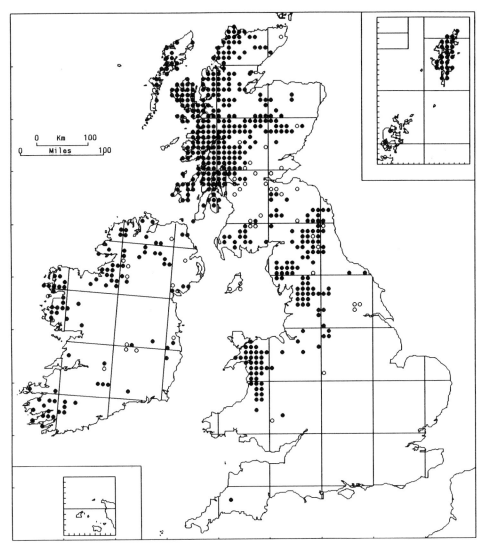

31/1. **Mylia taylorii** (Hook.) S. F. Gray

Forming large cushions in well-drained, sheltered situations, growing on rocks in block scree and woodland, on steep N.-facing slopes, and on stumps and the bases of trees in Highland birch woods. Common under *Calluna* and other dwarf shrubs in corries and screes; also on boggy moorlands and scattered among sphagnum. It is a typical component of the mixed montane liverwort communities known as the northern hepatic mat. 0–1230 m (Aonach Beag). GB 471+38*, IR 93+14*.

Dioecious. Capsules uncommon, ripe in late summer. Gemmae frequent.

From N. Scandinavia south to the Alps and Carpathians, most frequent and most luxuriant in the west. Azores, Himalaya, Japan, N. America, S. Greenland.

A. C. CRUNDWELL

143

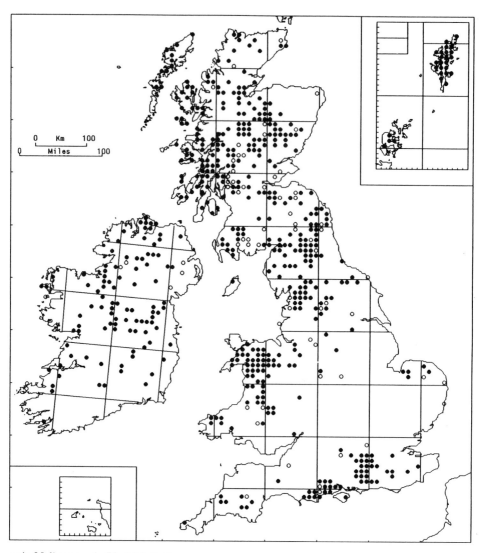

31/2. **Mylia anomala** (Hook.) S. F. Gray

Most commonly among sphagnum in bogs, but also found on wet heaths, on peaty rocks and in flushes, less frequently on rotting wood. 0–900 m (Glen Lyon). GB 436+60*, IR 97+8*.

Dioecious. Capsules very rare. Gemmae abundant.

Circumboreal. Common in N. Europe, extending south to the Pyrenees, N. Italy, Yugoslavia and Bulgaria.

A. C. CRUNDWELL

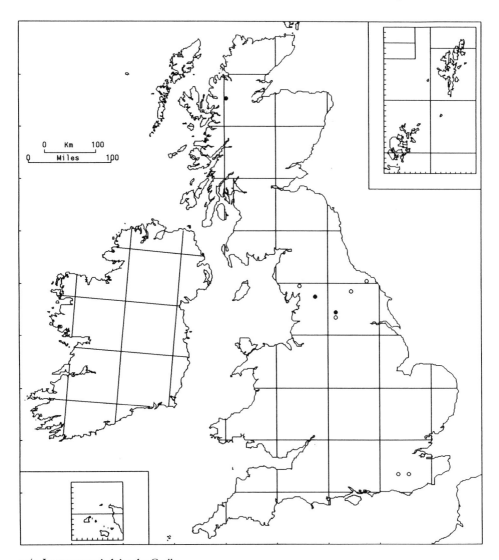

32/1. Jungermannia leiantha Grolle

A species of wet rocks and flushes in woodland and other shaded habitats, most of the records being from sheltered valleys or ravines. The rather wide variety of rock types includes sandstones, slates and shales, but always with some base content. Lowland. GB 3+6*.

Paroecious. Perianths and capsules are frequent, but gemmae are rare and have apparently been found only once in Britain.

Widespread in N. and C. Europe, and in the boreal zone of N. America. Not recorded from Asia.

The species is no longer known from many of the sites from which it was recorded in the nineteenth century, but the causes of its disappearance are obscure.

T. L. BLOCKEEL

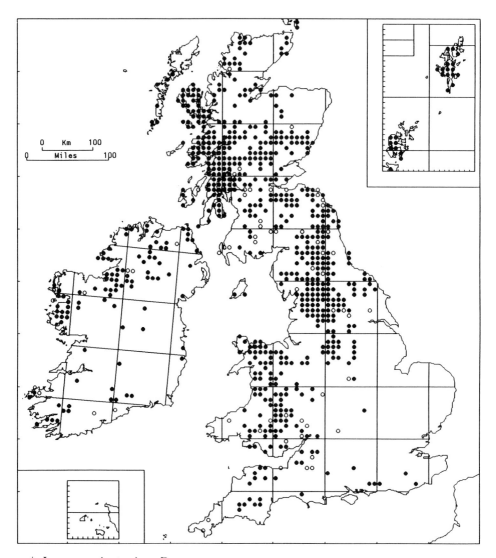

32/2. Jungermannia atrovirens Dum.

On moist or irrigated rocks and tufa in woodland, by streams, in ravines, on cliffs and in flushes, usually closely adherent to the rock or on a thin covering of soil. More rarely it occurs on shaded limestone walls, on damp, shaded soil, on clay banks and on gravelly ground by streams. It is characteristic of base-rich habitats on limestones and a wide range of other strata, but may occur also where the base content is low. 0–1070 m (Aonach Beag). GB 599+53*, IR 96+11*.

Dioecious. Perianths and male plants are frequent, and capsules are occasional, maturing in spring and summer. Gemmae are unknown.

Circumboreal. Throughout most of Europe but confined to mountains in the Mediterranean region.

T. L. BLOCKEEL

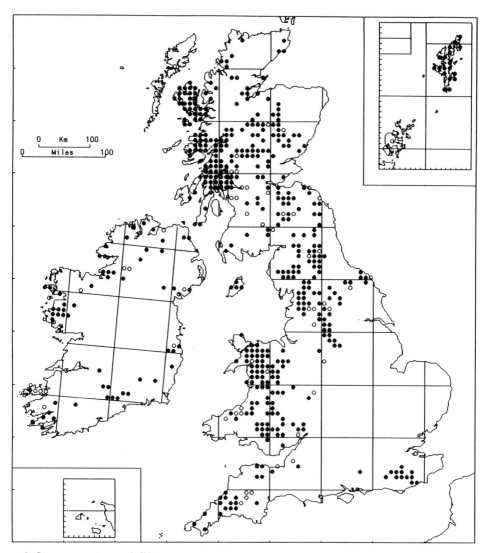

32/3. Jungermannia pumila With.

In thin appressed patches or mixed with other bryophytes in moist places on boulders and rock outcrops in wooded or sheltered situations, by streams and in gullies, and on cliffs. It may grow closely appressed to the rock or on thin soil on rocks and cliff ledges. More rarely it occurs in wet situations where the plants are often more robust and lax in habit, and on soil and gravelly ground kept moist by shade or ground-water. The substratum is commonly base-poor to moderately base-rich and the species is rare in strongly calcareous situations. 0–800 m (Snowdon). GB 420+44*, IR 57+14*.

Paroecious. Fertile plants and capsules are common; capsules mature in spring in the lowlands and in summer in the mountains. Gemmae are unknown.

Circumboreal. Throughout most of Europe but rare in the south.

T. L. Blockeel

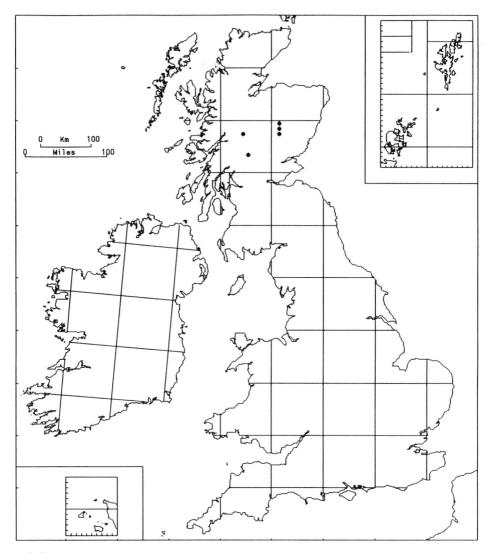

32/4. Jungermannia polaris Lindb.

In small appressed patches or mixed with other bryophytes on gravelly ground and on soil overlying base-rich rock, especially mica-schist and limestone, at high altitudes. 450–950 m (Coire Cheap, Ben Alder range). GB 5.

Paroecious. Fertile plants are frequent. Gemmae unknown.

Discontinuously circumpolar. Arctic and N. Europe; mountains of C. Europe.

T. L. BLOCKEEL

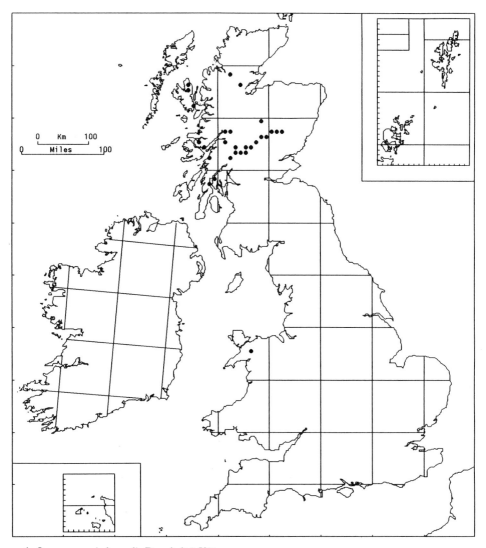

32/5. Jungermannia borealis Damsholt & Váňa

In small patches or creeping among other bryophytes on cliff ledges and in crevices of soft, moist base-rich rocks, usually with a northerly or easterly aspect. The species is confined to mountainous areas and the underlying rock formations include mica-schist and basalt. 100–1205 m (Ben Lawers). GB 28.

Dioecious. Perianths and male plants are frequent but capsules are rare, maturing in summer. Gemmae are unknown.

Arctic and N. Europe; mountains of C. Europe. Siberia, Canada (Hudson Bay), Greenland.

T. L. BLOCKEEL

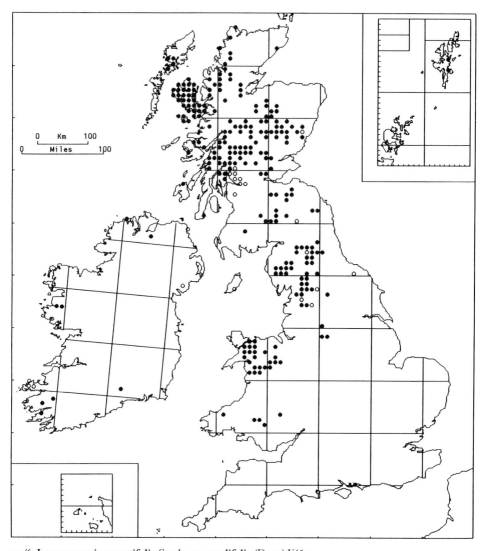

32/6. **Jungermannia exsertifolia** Steph. ssp. **cordifolia** (Dum.) Váňa

Usually in unmixed spongy masses on rocks and stones in flushes, runnels and streams, where the water is base-poor or moderately base-rich. Common associates in oligotrophic flushes include *Dicranella palustris* and *Philonotis fontana*. More rarely it occurs on moist or wet rock-faces, where the plants are usually less robust. 0–1100 m (Coire Ardair). GB 234+16*, IR 8+5*.

Dioecious. Perianths and male plants are occasional. Capsules are generally rare but more frequent at high altitudes. Gemmae are unknown.

Circumboreal. W. and C. Europe, north to the Arctic.

T. L. Blockeel

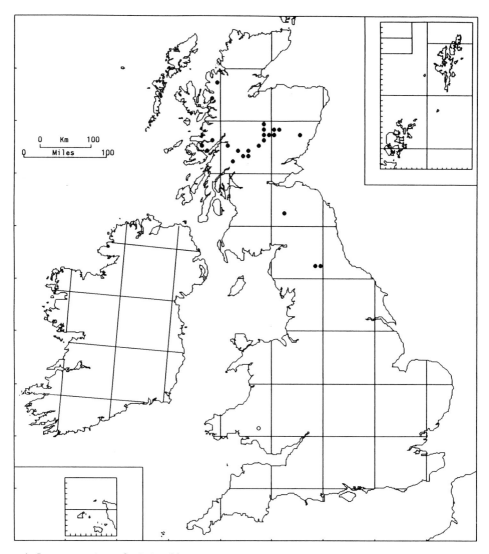

32/7. Jungermannia confertissima Nees

A species of moist or wet habitats in flushes, among rocks and on cliff ledges, and especially on the banks of streams and rivers, where it may be embedded in silty detritus. It thus occurs in similar habitats to the closely related *J. sphaerocarpa* but is apparently always a calcicole in the British Isles, occurring characteristically on limestones and schists. More rarely it occurs on drier ground, as in limestone quarries in Carmarthenshire and on mine waste in Argyll. 100–900 m (Glas Tulaichean). GB 24+1*.

Paroecious. Perianths and capsules are common, the latter maturing in spring and summer. Gemmae are unknown.

Widespread but rare in the mountainous regions of C. and N. Europe, north to the Arctic. Himalaya, boreal N. America (rare), Greenland.

T. L. BLOCKEEL

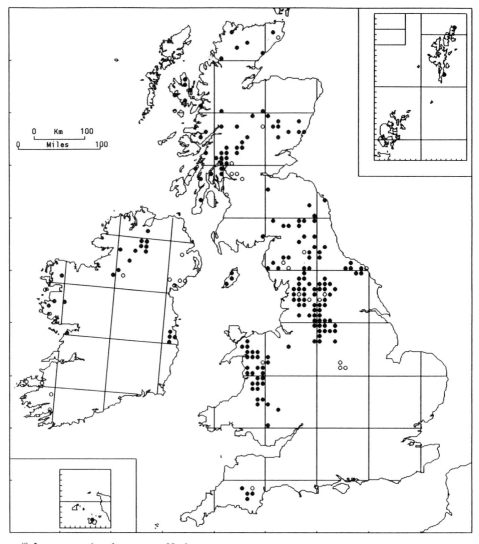

32/8. **Jungermannia sphaerocarpa** Hook.

On moist rocks, cliff ledges and stream banks, sometimes also on soil among rocks and on banks. The species may occur in very wet habitats and is tolerant of periodic submersion, often embedded in gritty detritus when growing by streams. The substratum varies from base-poor (often extremely so) to weakly base-rich and is frequently siliceous. 0–960 m (Ben Lawers). GB 173+22*, IR 20+12*.

Paroecious. Perianths are common and capsules frequent, maturing in spring and summer. Gemmae are unknown.

Circumboreal. Widespread in N., C., W. and E. Europe.

T. L. BLOCKEEL

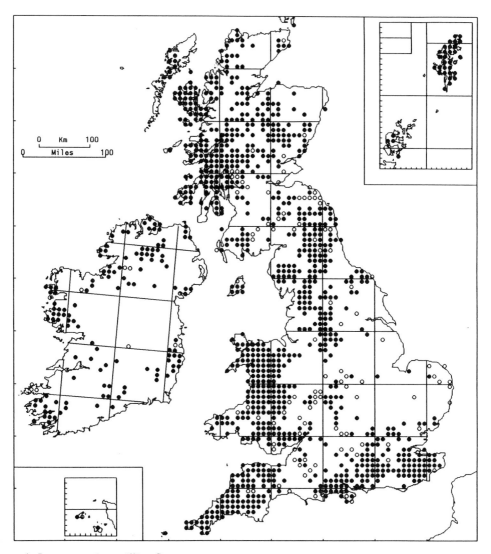

32/9. **Jungermannia gracillima** Sm.

A pioneer species of open or lightly shaded habitats on base-poor substrata, especially on clay, sand and loam. It may thus occur on banks and paths in woods and on moorland, on the banks of streams and ditches, by reservoirs and on cliffs, always in situations which are kept open by trampling, weathering or other disturbance. More rarely it occurs on soft rock such as porous sandstone and decaying shale on the banks of streams and in ravines. *Nardia scalaris* and *Dicranella heteromalla* are common associates. 0–800 m (Coire an t-Sneachda, Cairngorms). GB 940+106*, IR 125+11*.

Dioecious. Capsules are frequent, maturing in spring and summer. Gemmae are unknown.

Most of Europe but rare in the Mediterranean region. N. Africa, Asia, eastern N. America.

T. L. Blockeel

153

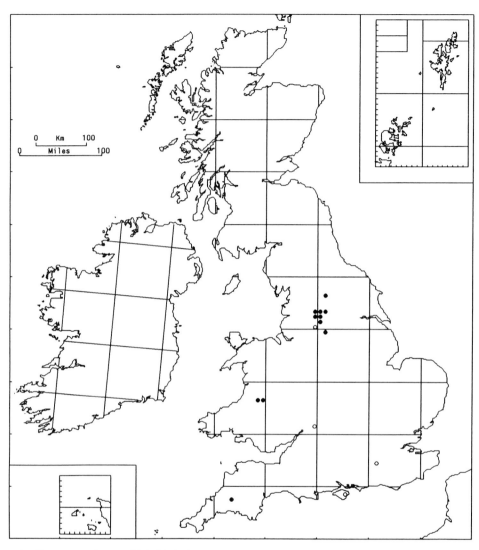

32/10. **Jungermannia caespiticia** Lindenb.

A pioneer species of moist base-poor soils in a variety of habitats, including moorland edges, stream banks, reservoir margins and mine workings. In the Pennines it favours clay soil overlying Millstone Grit and, more rarely, decaying shale, but it has also been recorded from sandstone rocks in southern England. It occurs in pure patches or in association with other bryophytes, characteristically *Calypogeia* spp., *Jungermannia gracillima* and *Nardia scalaris*, and is somewhat erratic and seasonal in occurrence. The main growing season is in the autumn. 0–300m (Peak District). GB 11+4*.

Dioecious. Perianths are rare and capsules are very rare in Britain. Gemmae are frequent.

Throughout much of C. and N. Europe. North America (Alaska along Bering Strait, New York State).

T. L. BLOCKEEL

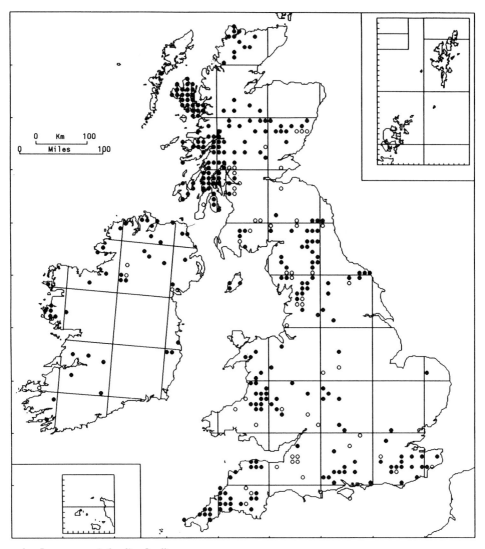

32/11. Jungermannia hyalina Lyell

Occurs most commonly on rocks near water-level by streams and in ravines, where it may be embedded in detritus, but also in a variety of other habitats such as moist gullies. It also occurs widely on sandy, loamy and clay banks in woods and by ditches, and on tracks and hollows kept moist by aspect or ground-water, and in these situations is widespread in lowland Britain. The substratum varies from acid to weakly basic. 0–300 m (Ben Cruachan). GB 263+53*, IR 40+6*.

Dioecious. Perianths are frequent, capsules occasional. Gemmae are unknown.

Circumboreal. Widespread in Europe but rare in the Mediterranean region.

Almost impossible to distinguish from *J. paroica* in the absence of gametangia; it is probably under-recorded in some areas and recorded erroneously in others.

T. L. Blockeel

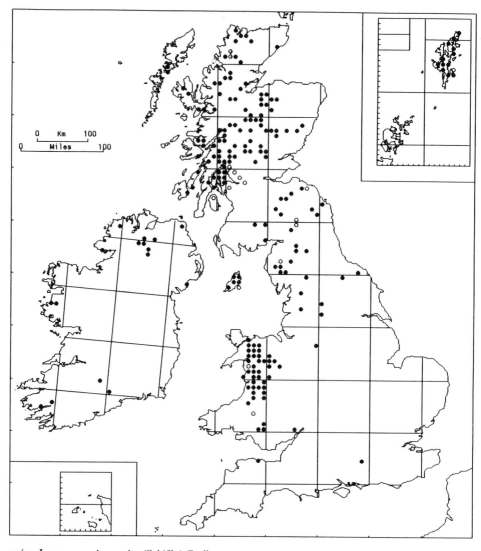

32/12. **Jungermannia paroica** (Schiffn.) Grolle

A species of wet, chiefly upland habitats on rocks in and by streams and on cliff ledges kept moist by dripping or seeping water. More rarely it occurs on irrigated gravelly banks by streams and in gullies, and on peaty ditch banks in the lowlands. Localities are base-poor to moderately base-rich and frequently shaded. 0–750 m (Glen Shiel). GB 184+16*, IR 16.

Paroecious. Perianths and capsules are common. Gemmae are unknown.

Endemic to Europe: W. Europe from Spain north to Faeroes and east to Belgium; very rare in C. Europe (Erzgebirge).

T. L. BLOCKEEL

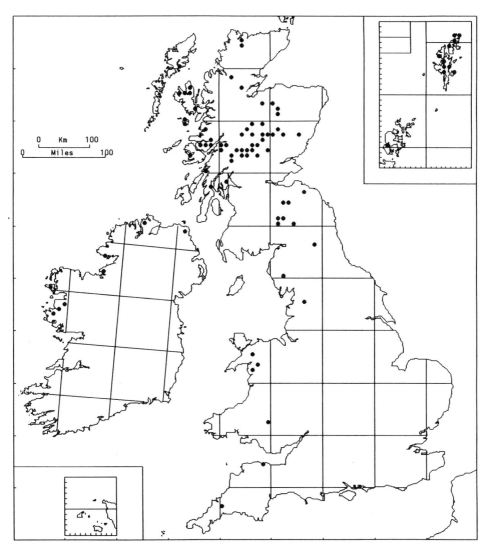

32/13. **Jungermannia subelliptica** (Lindb. ex Kaal.) Lev.

Most frequent in the upland districts of the British Isles, but at a variety of altitudes, in moist or periodically irrigated habitats in rock crevices, on cliff ledges, on disintegrating rock and on flushed gravelly banks and soil. It is usually in base-rich situations, as on mica-schist, but is not confined to them. 0–1050 m (Breadalbane). GB 79, IR 8.

Paroecious. Perianths and capsules are common. Gemmae are unknown.

Widespread in C. and N. Europe, especially in the mountains. N. America, Greenland.

Probably confused with *J. obovata* and therefore under-recorded.

T. L. BLOCKEEL

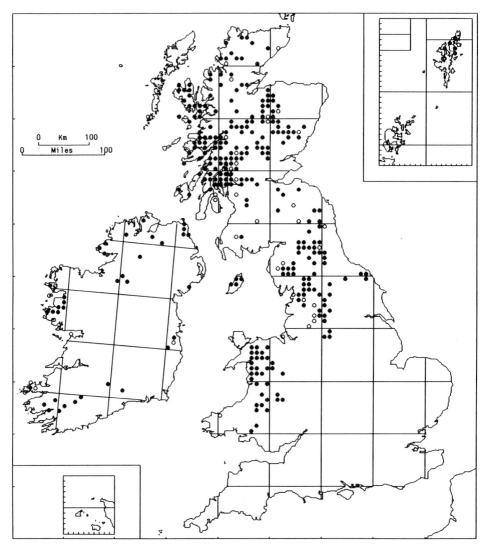

32/14. Jungermannia obovata Nees

This species is characteristic of rocks and boulders by streams, often growing in gullies and ravines where it is embedded in detritus near the water-level. It also occurs widely on dripping rocky banks and moist cliff ledges. The substratum varies from acid to moderately base-rich. 0–1100 m (Ben Alder). GB 257+34*, IR 36+6*.

Paroecious. Perianths and capsules are common. Gemmae are unknown.

Widespread in Europe, becoming montane southwards and absent from the Mediterranean region. Turkey, western and eastern N. America.

T. L. BLOCKEEL

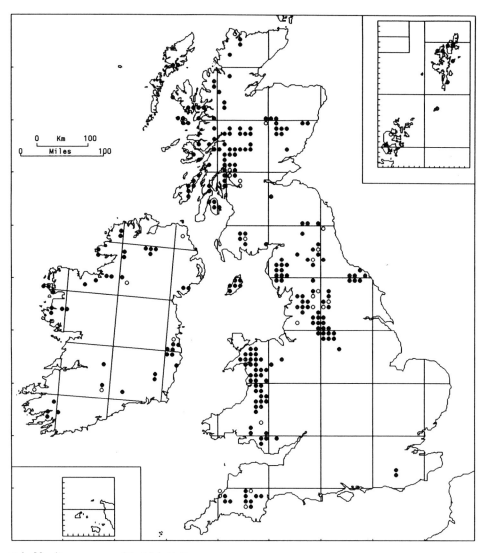

33/1. Nardia compressa (Hook.) S. F. Gray

On rocks in swift-flowing rivers and streams, on wet soil by streams, sometimes also in peaty pools and by the shores of lakes, where usually depauperate. 0–850 m (Braeriach, Cairngorms). GB 219+22*, IR 40+5*.

Dioecious. Capsules rare, ripe in April and May. Gemmae unknown.

In Norway and W. Sweden and the mountainous parts of C. and S. Europe. Turkey, E. Asia, western N. America, Greenland.

A. C. CRUNDWELL

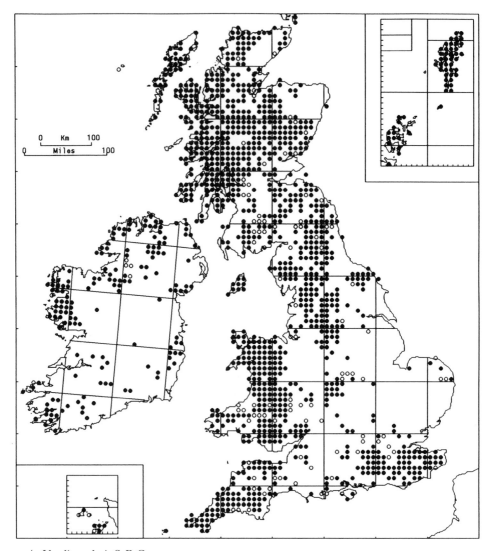

33/2. Nardia scalaris S. F. Gray

On damp base-poor sandy and peaty banks, on footpaths, damp roadsides, old quarries, edges of lakes and on wet rocks in streams. It is especially abundant on fine mineral soils at higher altitudes, growing on screes, gully floors and stony summit ridges. 0–1335 m (Ben Nevis). GB 1220+96*, IR 156+6*.

Dioecious. Capsules common, ripe in spring and early summer. Gemmae unknown.

Almost throughout Europe; in the south confined to the mountains. Macaronesia, Japan, N. America.

A. C. CRUNDWELL

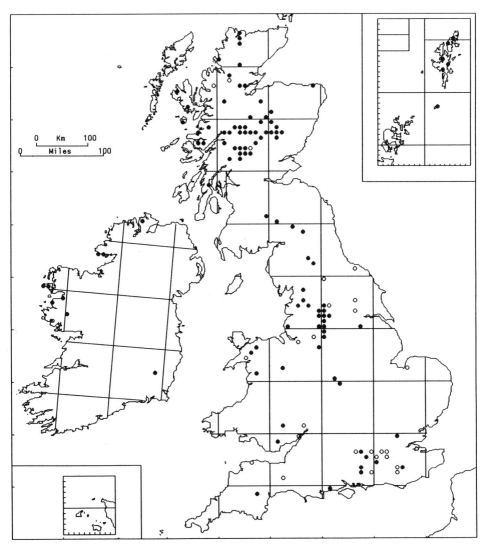

33/3. **Nardia geoscyphus** (De Not.) Lindb.

On moist basic soil on rock ledges and at the foot of mountain cliffs; also on sandstone rocks or on sandy or peaty soil in wet heaths or by water, often on banks. The habitat ranges from relatively dry to very wet, including streamsides and loch shores, sometimes submerged for most of the winter. 0–1070 m (Aonach Beag). GB 96+26*, IR 9.

Monoecious. Capsules common, ripe in spring and early summer. Gemmae unknown.

In Europe commonest in N. Scandinavia, extending southwards to the mountains of the Mediterranean region. Macaronesia, Caucasus, N. America, Greenland.

Probably under-recorded, because of its similarity to the very common *N. scalaris*.

A. C. CRUNDWELL

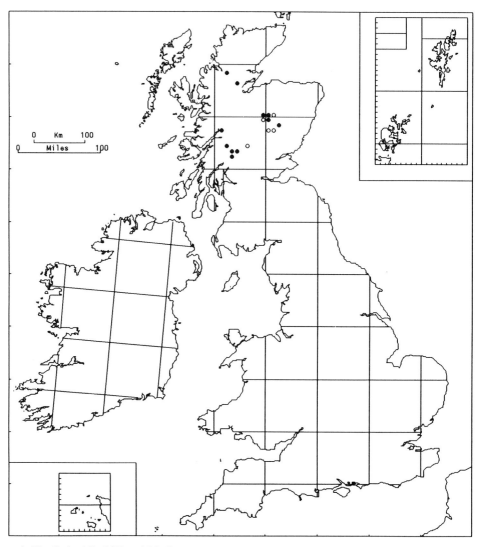

33/4. **Nardia breidleri** (Limpr.) Lindb.

On bare moist soil near the summits of some of the higher Scottish mountains. 850 m (Seana Bhraigh) to 1225 m (Aonach Beag). GB 11+5*.

Dioecious. Capsules frequent, ripe July and August. Gemmae unknown.

Circumpolar. On the mountains of Scandinavia and C. Europe.

A. C. CRUNDWELL

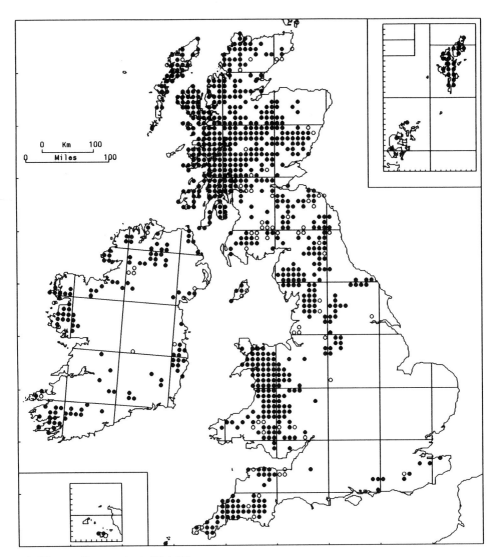

34/1. **Marsupella emarginata** (Ehrh.) Dum.

A frequent and often locally abundant species in a wide range of habitats. It most commonly grows in pure patches on rocks and detritus in and beside streams and lakes; also on soil-covered rocks, rock-walls in ravines, old quarries, stone walls, damp banks, gravelly footpaths, spoil heaps. and occasionally on boulders in woodlands. Substrates are typically acidic to neutral, rarely calcareous. Common associates are other calcifuges such as *Diplophyllum albicans*, *Polytrichum juniperinum* and *P. piliferum*. On mountains it grows on open stony slopes, rocks by streams, stony flushes, flushed rock-slabs and beside snow-beds. 0–1220 m (Ben Nevis). GB 731+68*, IR 118+10*.

Dioecious; fertile plants are common, sporophytes occasional.

A widespread temperate circumboreal and disjunct tropical-montane species. Mountains of Europe south to Iberian Peninsula and Turkey, north to Scandinavia, Faeroes and Iceland. Macaronesia, tropical African mountains, Siberia, Himalaya, E. Asia (widespread), N. America, Greenland, Mexico, Colombia.

The three varieties are poorly recorded and cannot be mapped separately.

D. G. LONG

163

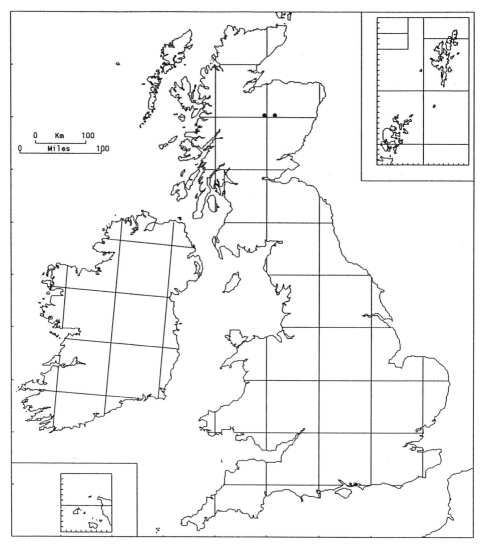

34/1A. **Marsupella arctica** (Berggr.) Bryhn & Kaal.

One locality is close to a small pool in a deep U-shaped mountain valley, where the species occurs in relatively pure tufts in moist sheltered declivities of a mossy W.-facing block-scree. Its other station is in a late-snow patch in an incised section of a small stream, where it grows mixed with *M. sphacelata* in the vicinity of several typical snow-bed mosses. 810 m and 1120 m. GB 2.

Dioecious; gametangia not found in Scotland.

Svalbard. Soviet Far East, N. America, Greenland.

First discovered in Scotland in 1989. The notes given above are based on a paper by Long *et al.* (1990), to which the reader is referred for further details.

M. O. HILL

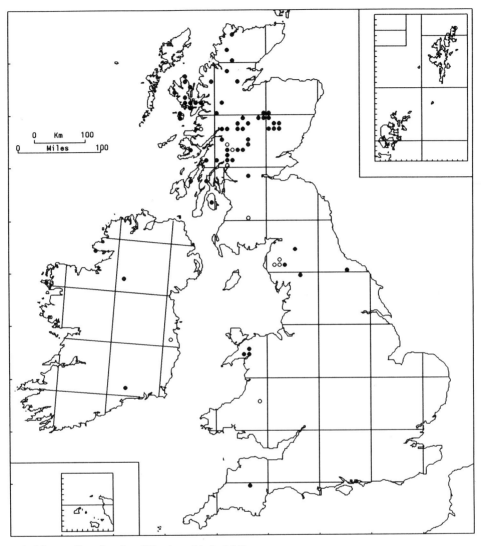

34/2. **Marsupella sphacelata** (Gieseke ex Lindenb.) Dum.

A calcifuge common only in montane habitats such as late-snow areas and sheltered N.- and E.-facing gullies, where it grows on wet rock-slabs and boulders in and by streams and flushes. Plants on more exposed rocks and gravelly or peaty slopes tend to be smaller in stature and grow in mixed patches with other dwarf species such as *Anthelia julacea*, *Lophozia sudetica*, *Polytrichum piliferum* and *Racomitrium* spp. Plants growing in springs, on wet rocks and submerged in snow-melt streams and lake margins frequently grow in robust masses, often pure or mixed with *Scapania uliginosa*, *S. undulata*, *Philonotis fontana* etc. At lower altitudes the plant is rare though widespread. In Argyll, and perhaps in Wales, it is associated with copper-rich rocks. Sea-level (Wales, Shetland) to 1225 m (Aonach Beag); rare below 500 m. GB 61+11*, IR 4+1*.

Dioecious; fertile plants and sporophytes occasional.

Circumpolar. Mountains of Europe south to Greece and Spain, north to Iceland and Scandinavia. Azores, Japan, N. America, Greenland.

D. G. Long

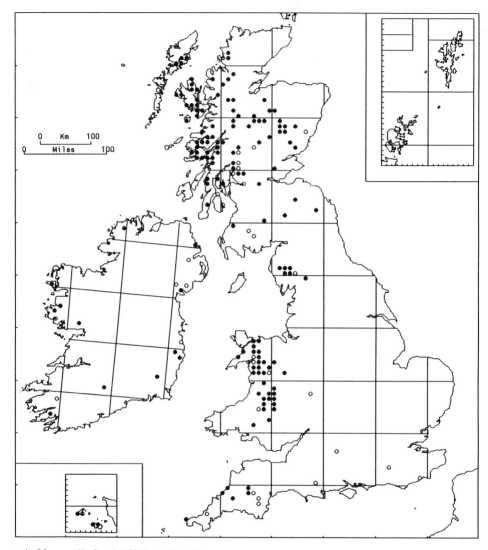

34/3. **Marsupella funckii** (Web. & Mohr) Dum.

A more 'weedy' species than the other dwarf *Marsupella* species, this plant is most often seen forming prostrate blackish patches on trampled roadsides and footpaths in north and west Britain and Ireland. These patches may be pure or mixed with *Diplophyllum albicans*, *Nardia scalaris* and a dwarf form of *M. emarginata* which mimics *M. funckii* in appearance. It is also recorded from a range of less typical habitats such as cliff ledges, soil-filled crevices on rocks by streams, soil-capped boulders, wall tops and mine waste. Like most other *Marsupella* species it is strictly calcifuge, but unlike them it is absent from the high mountains. 0–670 m (Berwyn Mts). GB 132+28*, IR 12+5*.

Dioecious; fertile plants common, sporophytes rare.

Found throughout the subalpine parts of C. Europe, south to Turkey and Portugal, north to S. Scandinavia, Faeroes. Macaronesia, N. America (Appalachians). Absent from the Arctic.

D. G. Long

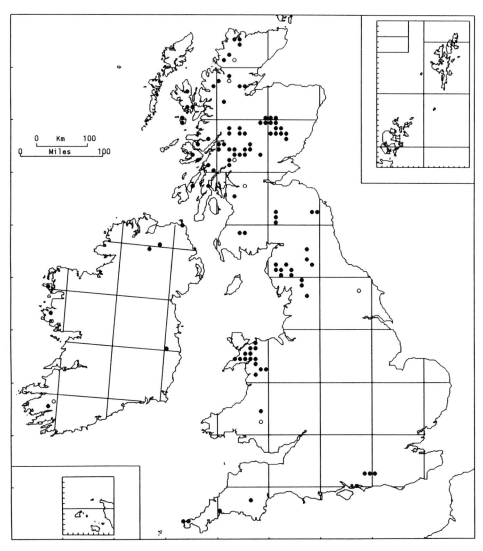

34/4. **Marsupella sprucei** (Limpr.) H. Bern.

Like *M. adusta* it typically grows in small erect tufts on small siliceous stones and boulders – especially granite and gneiss – in screes and fell-fields on the higher hills and mountains. These tufts are usually pure. It can also grow on larger boulders, in rocky gullies, on partly buried stones in springs and flushes, around snow beds, on low cliffs and rarely on exposed peaty or gravelly soil. In such habitats the plants are often mixed with dwarf bryophytes such as *Cephalozia bicuspidata*, *Gymnomitrion concinnatum*, *Lophozia sudetica*, *Nardia scalaris*, *Scapania scandica*, *Ditrichum zonatum* and *Kiaeria starkei*. Sea-level (Moidart) to 1335 m (Ben Nevis). GB 100+6*, IR 7+1*.

Paroecious; gametangia and sporophytes common.

A bipolar arctic-alpine. Mountains of Europe south to Spain, north to Scandinavia and Iceland. Madeira, N. America, Greenland, disjunct in the Southern Hemisphere in Argentina and New Zealand.

D. G. LONG

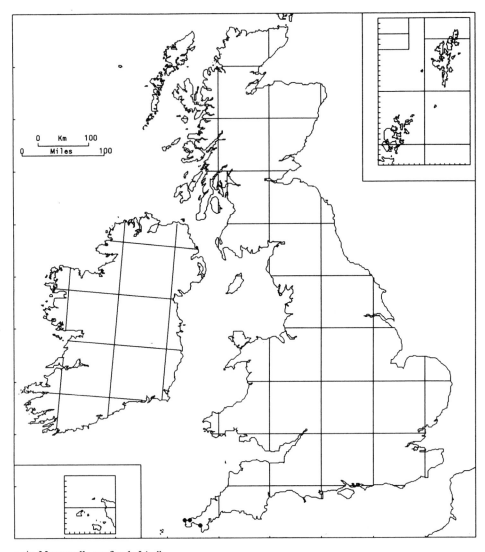

34/5. **Marsupella profunda** Lindb.

Known only from disused china-clay works, where it grows as a colonist of moist, crumbling mica-rich granite and of the clay formed by decomposition of this rock. Associated species include *Cephalozia bicuspidata*, *Cephaloziella divaricata*, *Diplophyllum albicans*, *Marsupella sprucei*, *Nardia scalaris*, *Scapania compacta* and *S. irrigua*. Lowland. GB 3.

Paroecious; gametangia and sporophytes common.

A Lusitanian species, known outside the British Isles only from Portugal and Madeira, but perhaps overlooked elsewhere in W. Europe.

M. profunda is closely related to *M. sprucei*. It has only recently been recognized in Britain; these notes are taken from Paton's (1990a) report of its discovery. The species was found at four sites in Cornwall between 1965 and 1971. It has not been seen since 1971, and the localities in which it grew have been destroyed or become overgrown with gorse or bramble. However, it doubtless survives elsewhere in the area.

C. D. Preston

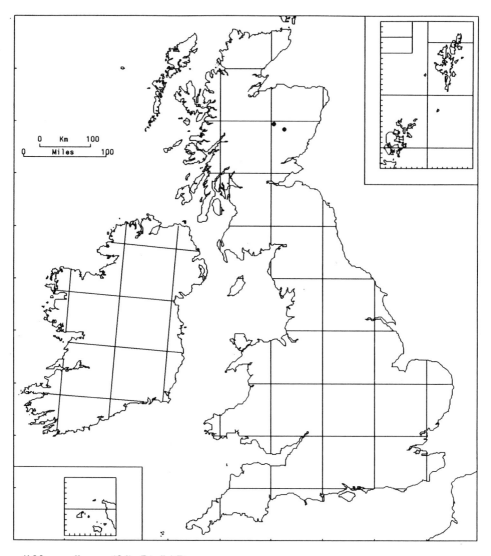

34/6 **Marsupella sparsifolia** (Lindb.) Dum.

It grows in low blackish patches on wet acidic rocks, especially granite, in and beside snow-melt streams in areas of late snow-lie. It is commonly associated with *M. sphacelata*, with which it is readily confused in the field. 950 m (Lochnagar) to 1030 m (Beinn a' Bhuird). GB 2.

Paroecious; gametangia and sporophytes usually present.

A local bipolar arctic-alpine species (probably under-recorded) in the higher mountains of Europe, more widespread and at lower altitudes in Scandinavia. Also in the Azores, N. America, Greenland, E. Africa (Uganda), S. Africa, New Zealand.

D. G. LONG

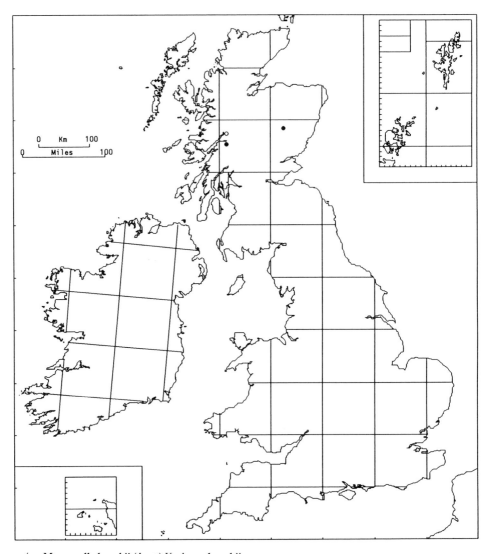

34/7a. **Marsupella boeckii** (Aust.) Kaal. var. **boeckii**

A mountain plant growing in thin pale green often pure prostrate patches on wet acidic rock walls and large blocks, associated with *Diplophyllum albicans*, *Marsupella emarginata* etc.; in one site at lower altitude mixed with a filamentous alga, *Scapania undulata* and *Calliergon sarmentosum*. 700–1090 m (Ben Nevis). GB 2+1*.

Dioecious; gametangia often present, sporophytes rare.

A rare arctic-alpine, reported in Europe from the Pyrenees, Alps, Carpathians, Scandinavia, Faeroes. Elsewhere, in north-east U.S.A., Greenland, Bhutan.

D. G. LONG

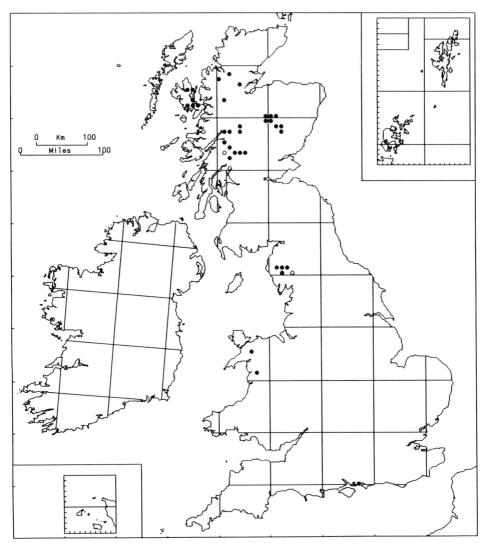

34/7b. **Marsupella boeckii** (Aust.) Kaal. var. **stableri** (Spruce) Schust. (*M. stableri* Spruce)

Grows in distinctive purplish-brown or coppery mats on periodically flushed montane acid rock-walls, boulders and gravelly soil, usually on N.- to E.-facing slopes, and often in gullies and late-snow areas. Occasionally on basic rocks such as basalt. On rocks the patches are often pure but frequent associates include *Anthelia julacea*, *Diplophyllum albicans* and *Gymnomitrion obtusum*. In exposed stony or gravelly sites, populations are more commonly mixed with other small bryophytes such as *Anthelia juratzkana*, *Cephalozia bicuspidata*, *Lophozia sudetica*, *Nardia scalaris*, *Ditrichum zonatum* and *Oligotrichum hercynicum*. Commonly between 400 m and 1200 m (Ben Nevis). GB 34+4*.

Dioecious; fertile plants are frequent, sporophytes very rare.

Outside the British Isles its distribution is uncertain but it has been reported from Canada (British Columbia). J. A. Paton (pers. comm.) is of the opinion that it is worthy of specific rank.

D. G. LONG

171

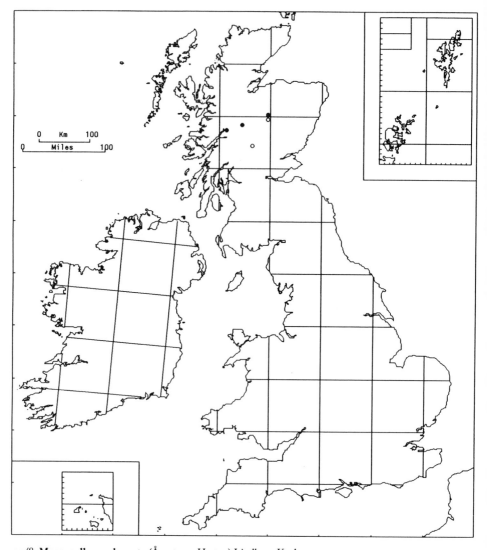

34/8. **Marsupella condensata** (Ångstr. ex Hartm.) Lindb. ex Kaal.

In Scotland an exclusively alpine species reported from only a few of the highest mountains where it grows on moist peaty or gravelly soil on bare stony summits and exposed slopes, often in or close to late-snow areas. Although occasionally in pure patches, the plants are usually mixed with other dwarf bryophytes characteristic of such extreme habitats such as *Gymnomitrion concinnatum*, *Marsupella* spp., especially *M. brevissima*, *Nardia breidleri*, *Scapania scandica*, *Conostomum tetragonum*, *Ditrichum zonatum* and *Kiaeria starkei*. In such mixed mats it is hard to detect. 980–1240 m (Ben Nevis). GB 3+2*.

Dioecious; female plants common, male plants and sporophytes not found in Britain.

Arctic-alpine: Alps, Scandinavia, Iceland, Greenland.

D. G. Long

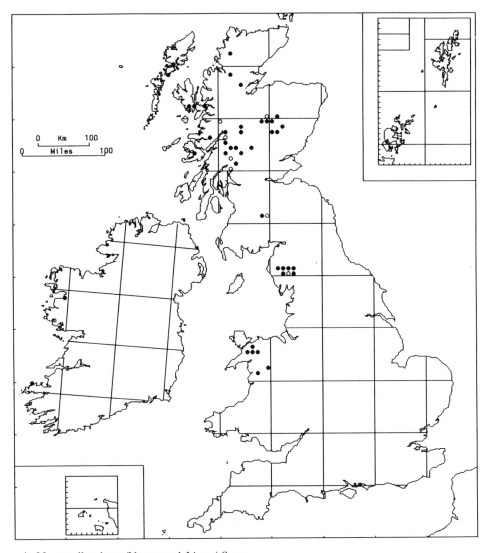

34/9. **Marsupella adusta** (Nees emend. Limpr.) Spruce

An alpine species of similar size to *M. sprucei* and occurring in similar habitats in the higher mountains, including screes, exposed fell-fields, ridges, summits, streamsides, flushes and loch margins. Small stones of granite or gneiss are the favoured substrate, more rarely basic schists or basalt. It also grows less typically on moist rock-walls and slabs and rarely on moist gravel soil amongst rocks. Apart from *M. sprucei*, associates include other dwarf *Marsupella* species, *Diplophyllum albicans* and *Gymnomitrion* species. Sea-level (Moidart) to 1200 m (Ben Lawers), but rare at low altitudes and mostly above 400 m. GB 40+7*, IR 2.

Paroecious; fertile plants and sporophytes common.

Suboceanic-montane. Scattered through the Alps and other C. and E. European mountains, Norway. Azores, Madeira. Absent from N. America.

D. G. LONG

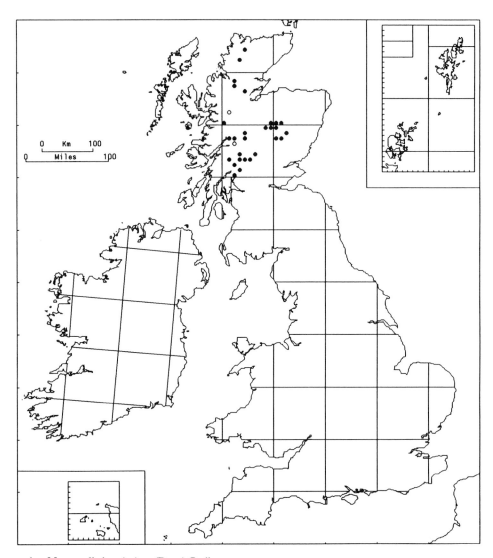

34/10. Marsupella brevissima (Dum.) Grolle

Of all the dwarf alpine *Marsupella* species, *M. brevissima* is often the easiest to detect in the field as it frequently forms extensive blackish mats on peaty or gravelly soil on mountain summits, exposed fell-fields, moraines, acid screes and around late-snow beds and springs. It commonly, but less conspicuously, grows in mixed mats in these habitats, with a range of other dwarf species such as *Anthelia juratzkana*, *Diplophyllum albicans*, *Gymnomitrion* spp., other *Marsupella* spp., *Nardia breidleri*, *N. scalaris*, *Scapania scandica*, *Conostomum tetragonum*, *Ditrichum zonatum*, *Kiaeria* spp., *Oligotrichum hercynicum*, *Polytrichum juniperinum* and *Racomitrium* spp. 440 m (Seana Bhraigh) to 1340 m (Ben Nevis), but rare below 900 m. GB 29+2*.

Paroecious and autoecious, often fertile; sporophytes common.

Arctic-alpine. C. and E. European mountains, Scandinavia, Faeroes, Iceland. E. Himalaya, western N. America, Greenland.

D. G. LONG

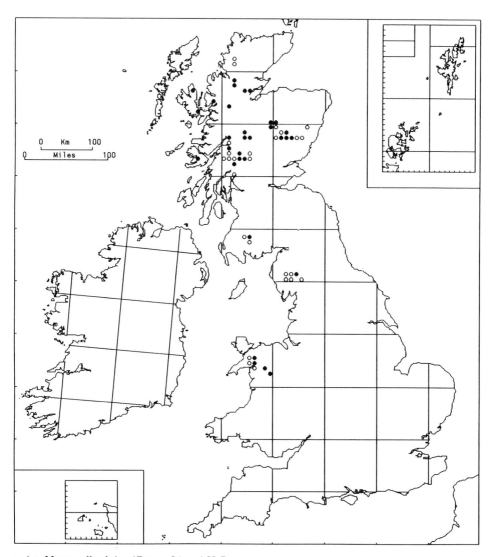

34/11. **Marsupella alpina** (Gott. ex Limpr.) H. Bern.

This species favours wet or intermittently irrigated rock-walls, crevices in gullies on cliffs, slabs, dripping rocks and damp scree blocks, usually in sites with a sheltered north to east aspect on mountains. Granitic and gneissic rocks are its usual substrates, but it can grow on more basic schists and basalt. It occurs more rarely on peaty or gravelly soil, but almost only in damp rocky sites close to flushes or snow patches. The *Andreaea*-like patches, though not extensive, are often relatively pure; recorded associates are few, including *Anthelia julacea*, *Gymnomitrion crenulatum*, *G. obtusum*, *Marsupella emarginata* and *Ditrichum zonatum*. 300 m (Bidean nam Bian) to 1340 m (Ben Nevis), but common only above 650 m. GB 31+26*.

Dioecious; fertile plants common, sporophytes occasional.

Suboceanic-alpine. Mountains of C. and E. Europe, south to Iberian Peninsula; at lower altitudes in Scandinavia. Nepal, Japan, western N. America.

D. G. LONG

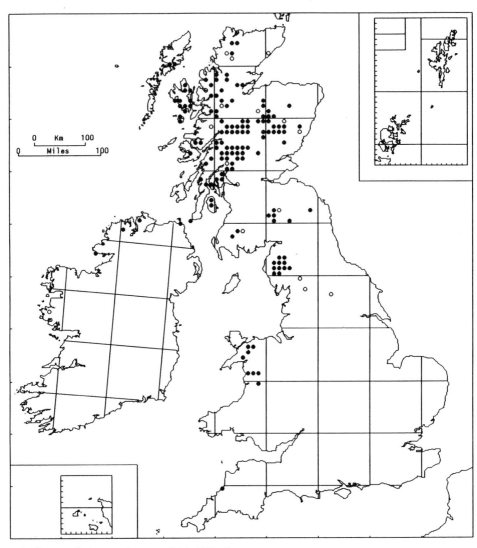

35/1. **Gymnomitrion concinnatum** (Lightf.) Corda

A montane species forming distinctive yellow-green to greyish mats or cushions on siliceous rocks, gravel and mineral soils in exposed to sheltered situations, including cliff ledges, acid screes, fell-fields and rocky gullies. It is a common component of the vegetation of exposed summits, moraines and late-snow areas, often in mixed bryophyte turfs with other *Gymnomitrion* spp., *Diplophyllum albicans, Lophozia sudetica, Nardia scalaris, Marsupella* spp. (especially *M. brevissima*), *Ditrichum zonatum, Kiaeria starkei, Polytrichum juniperinum, P. piliferum, Racomitrium* spp. and lichens. Occasionally it grows on basic schists, basalt and limestones, but usually where a leached peaty crust has developed. Found sporadically at low altitudes in the north and west from sea-level (Shetland) to 300 m, but mainly above this level, to 1340 m (Ben Nevis). GB 132 + 18*, IR 4 + 1*.

Dioecious; commonly with gametangia, sporophytes occasional.

Circumboreal arctic-alpine. European mountains south to Turkey; Scandinavia, Faeroes, Iceland, Svalbard. Siberia, Japan, N. America, Greenland.

D. G. Long

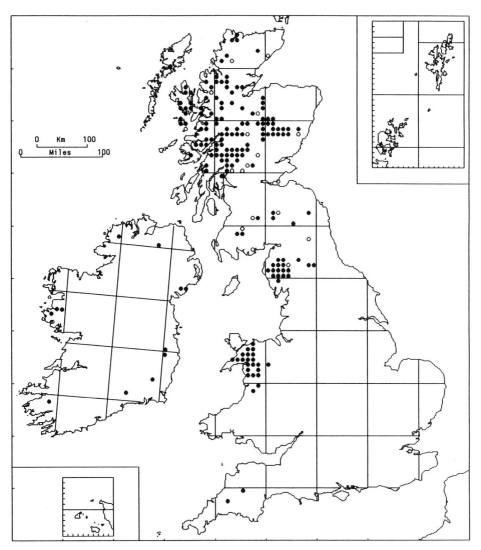

35/2. **Gymnomitrion obtusum** (Lindb.) Pears.

A conspicuous hepatic growing in whitish or greyish-green cushions or mats on both siliceous and basic rocks, but rarely on soil. Rock habitats include ledges and tops of large boulders, block-scree, cliff ledges, crumbling slopes in gullies, and stones in and around snow patches. It grows with species such as *G. concinnatum*, *G. crenulatum*, *Diplophyllum albicans*, *Lophozia sudetica*, *Marsupella* spp., *Andreaea* spp., *Racomitrium heterostichum*, *R. lanuginosum* and xerophytic lichens. On soil it is an occasional component of dwarf bryophyte crusts on exposed summits and ridges, but rarer in this habitat than *G. concinnatum*. In the north and west it can descend (rarely) to sea-level, but is common only between 500 m and 1200 m (Ben Nevis). GB 160+18*, IR 12+1*.

Dioecious; commonly fertile, sporophytes common.

Its oceanic subarctic-alpine distribution is similar to that of *G. concinnatum* but more restricted to high-rainfall regions; it is rare in the Arctic. European mountains south to Spain, Portugal and Caucasus, north to Scandinavia. Western N. America, S. Greenland. Disjunct in E. Himalaya (Bhutan).

D. G. LONG

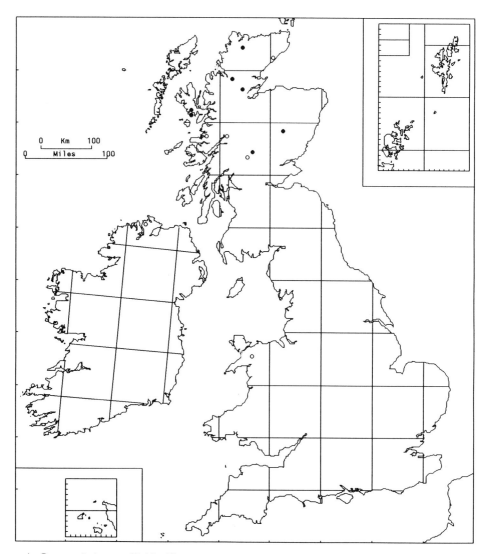

35/3. Gymnomitrion corallioides Nees

Although a much rarer species than *G. concinnatum* and *G. obtusum*, it appears to have a broader ecological tolerance, occurring on soil and on a wider spectrum of rock types from acid granite and gneiss to calcareous schist; on this last substrate, e.g. on the Breadalbane mountains and Ben Hope, it can grow in some abundance. It particularly favours small, exposed, often unstable rock outcrops, sloping slabs, low cliffs and large blocks, where it can form extensive patches. Other habitats include peaty and gravelly mineral soils on exposed summits and in late-snow areas where it is an occasional member of dwarf bryophyte mats. On rock, patches are often pure and close associates few, e.g. *Gymnomitrion concinnatum, G. obtusum, Lophozia sudetica, Pohlia nutans, Racomitrium lanuginosum* and lichens. 400 m (Ben Hope) to 1205 m (Ben Lawers). GB 7+5*, IR 2*.

Dioecious; fertile plants common, sporophytes occasional.

A circumpolar arctic-alpine, found throughout the high mountains of central Europe, to Faeroes, Iceland, Scandinavia, Svalbard. Siberia, Japan, N. America, Greenland. Especially abundant in the Arctic.

According to Müller (1954–57), of all European hepatics, it grows at the highest altitude (Alps).

D. G. Long

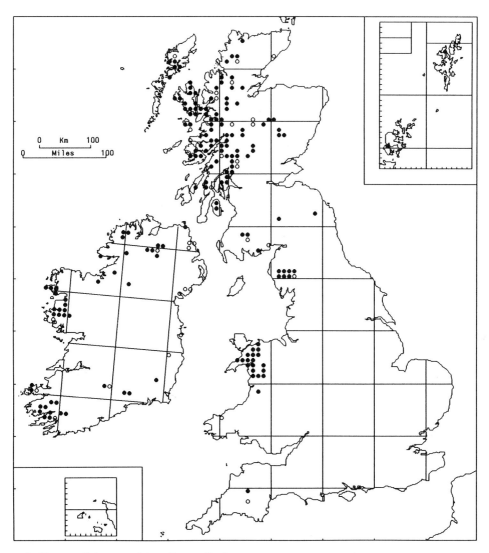

35/4. Gymnomitrion crenulatum Gott. ex Carringt.

Ecologically similar to other British *Gymnomitrion* species, it occurs on a wide range of rock types, including quartzite, gneiss, sandstone, schists and basalt. It grows in small often inconspicuous low patches or rounded cushions as a pioneer of exposed sites such as moist to relatively dry rock-walls, rocky gullies, boulders and screes. These sites are often sun-exposed. Common associates are *Diplophyllum albicans*, *Gymnomitrion obtusum*, *Andreaea rothii*, *A. rupestris* and several lichens. It does not grow on soil. In W. Scotland it ranges from sea-level (Loch Hourn) to 700 m (Glen Duror), but reaches 920 m in Ireland (Galtee Mts). GB 123 + 22*, IR 43 + 13*.

Dioecious; fertile plants common, sporophytes occasional.

A truly North Atlantic species endemic to Europe, occurring in the regions close to the western seaboard from Spain and Portugal to S.W. Norway.

D. G. LONG

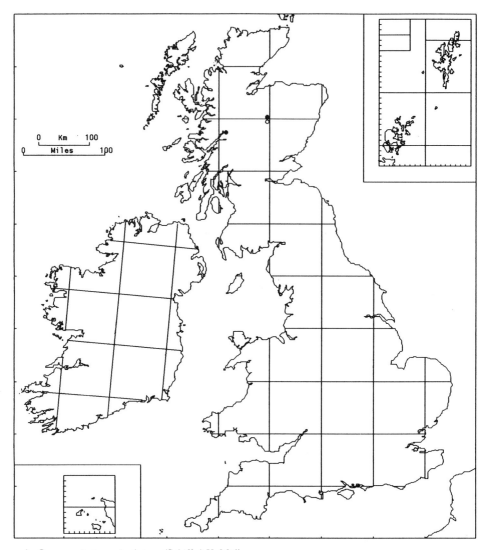

35/5. **Gymnomitrion apiculatum** (Schiffn.) K. Müll.

A dwarf alpine hepatic growing on peaty or gravelly mineral soils on gently sloping exposed N.- to E.-facing slopes and moist depressions in open fell-fields, often in areas of late snow-lie. It is easily overlooked as it is often mixed in dwarf bryophyte mats or lichen-rich crusts with *Anthelia juratzkana, Cephalozia bicuspidata, Gymnomitrion concinnatum, Marsupella brevissima* etc. 1100–1340 m (Ben Nevis). GB 2+1*.

   Dioecious; only female plants known in Scotland.

   An arctic-alpine found in the Alps and Tatra Mts, Scandinavia, north-west U.S.S.R., Svalbard. Japan, Alaska, Greenland.

D. G. LONG

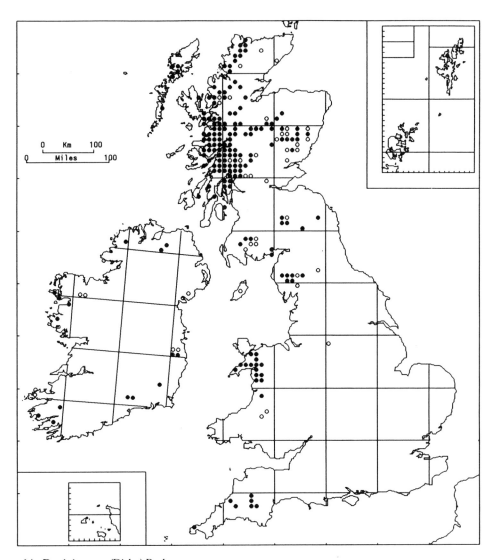

36/1. **Douinia ovata** (Dicks.) Buch

It grows as close patches on trees and rocks. *Betula* woods and wooded ravines are especially favoured, where it grows on trunks and larger limbs of trees, as well as mossy boulders and rarely peaty banks. It also grows on *Quercus* and *Sorbus aucuparia*. On more open slopes (possibly often where woodland formerly occurred) it grows on mossy boulders in large block screes. Rocks are often acidic but it has been found on basalt and basic schist. On both rock and wood it grows with a similar range of associates, such as *Diplophyllum albicans*, *Frullania tamarisci*, *Plagiochila punctata*, *Scapania gracilis*, *Dicranum fuscescens*, *D. scoparium* and *Hypnum cupressiforme*, all of which can tolerate intermittent drought. Sea-level (W. Highlands and Skye) to 550 m, rarely at higher altitudes; to 900 m on Ben Lawers. GB 164+37*, IR 15+10*.

Dioecious; fertile plants and sporophytes common. Gemmae absent.

In Europe a distinctly oceanic taxon in regions close to the Atlantic coasts from Portugal to the Faeroes and Scandinavia; more locally east to Finland, Germany and E. France. Disjunct in high-rainfall areas of Japan, Pacific N. America and S. Greenland.

D. G. Long

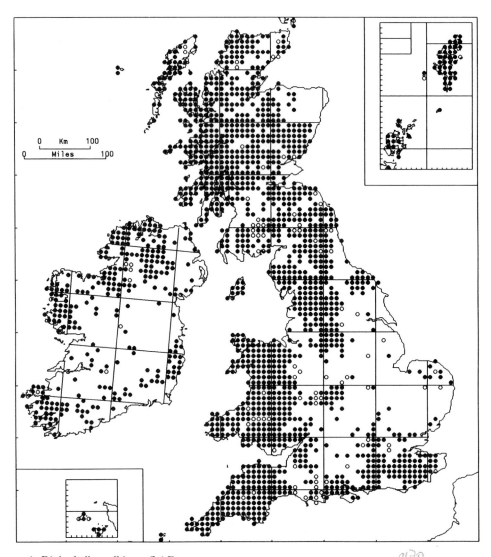

37/1. **Diplophyllum albicans** (L.) Dum.

A calcifuge occurring in a wide range of habitats such as soil, stumps, logs, boulders and rock-faces in woodlands and wooded ravines, banks by footpaths and forestry roads, streamsides, rocky gullies, sphagnum bogs, heaths and, in mountains, in virtually all habitats from cliff ledges, block-screes, streamsides and moraines to exposed rocky summits and fell-fields. Although a calcifuge, it can grow in calcareous areas on thin leached soil, humus or mats of other bryophytes. It is also a common component of montane and oceanic-montane bryophyte and hepatic mats. 0–1340m (Ben Nevis). GB 1435+84*, IR 270+10*.

Dioecious; perianths and sporophytes frequent. Gemmae commonly produced but not abundant.

A widespread circumboreal-suboceanic species. Europe south to Iberian Peninsula and Turkey, north to Faeroes, Scandinavia, Svalbard, Iceland. Macaronesia, Siberia, Kamchatka, Japan, Korea, Taiwan, Hawaii, N. America, Greenland.

D. G. LONG

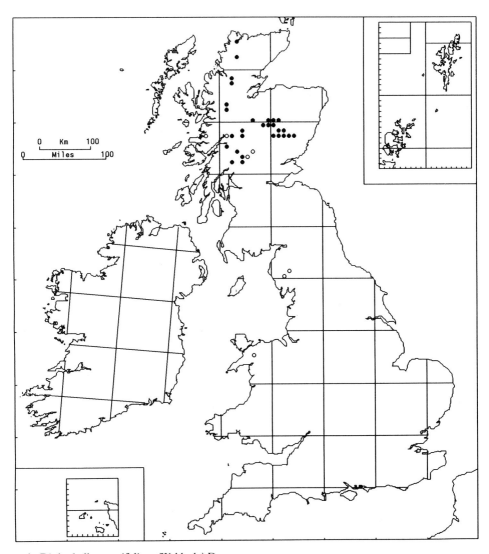

37/2. Diplophyllum taxifolium (Wahlenb.) Dum.

A montane and alpine calcifuge found only on the higher hills and mountains, usually growing on rocky or peaty banks, hollows in rocky fell-fields and on boulders and rock ledges, either in pure tufts or with other calcifuge hepatics such as *Diplophyllum albicans*, *Gymnomitrion concinnatum*, *G. obtusum*, *Lophozia sudetica* and *Tetralophozia setiformis*. It is often associated with late-snow areas, especially on shaded moist rocks. 520–1130m (Ben Lui). GB 28+7*.

Dioecious; gametangia rare, sporophytes unknown in Britain. Gemmae common.

A circumpolar subarctic-alpine. Mountain ranges of Europe from Turkey and Pyrenees (rare) north to Scandinavia and Svalbard, becoming commoner northwards. Siberia, Kamchatka, Japan, Sakhalin, Taiwan, N. America, Greenland.

It is probably often overlooked in the field because of confusion with the ubiquitous *D. albicans*.

D. G. Long

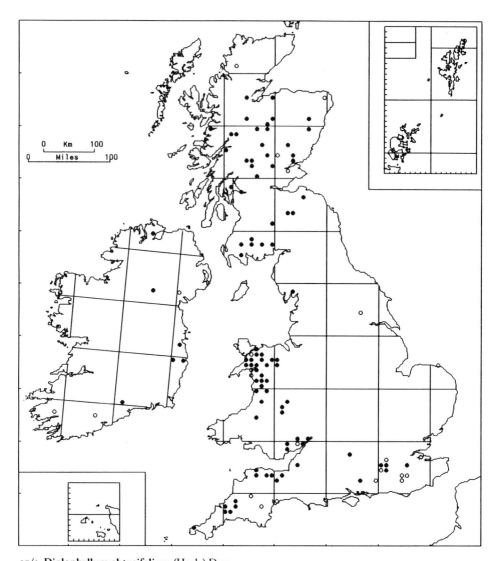

37/3. **Diplophyllum obtusifolium** (Hook.) Dum.

A calcifuge pioneer of disturbed soil in old quarries and gravel-pits and on streamsides and by roads. In forestry plantations it is often found by unmetalled roads and tracks, where it grows on damp crumbling banks, gravelly slopes and ditch sides, mixed with other small colonists such as *Diplophyllum albicans*, *Nardia scalaris*, *Scapania scandica*, *Ceratodon purpureus*, *Ditrichum heteromallum* and *Pogonatum urnigerum*. 70–350 m. GB 86+21*, IR 6+4*.

Paroecious; perianths and sporophytes frequent. Gemmae rare.

Throughout Europe but rare in extreme south and north. Japan, Taiwan, north-west U.S.A., Greenland.

It may be extending its range in Scotland in response to the increase in commercial forestry and proliferation of forestry roads in recent years. It is probably under-recorded, as it is easily overlooked in the field.

D. G. Long

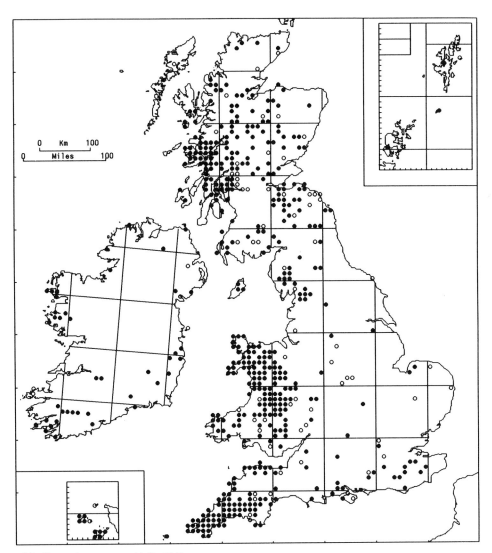

38/1. Scapania compacta (A. Roth) Dum.

An aptly named *Scapania* growing in low dense tufts or extensive sheets in more xeric habitats than most other members of the genus, on exposed dry rock outcrops and boulders, and in rocky gullies, ravines and quarries, especially on rocks with a thin cover of or pockets of sandy soil or detritus. It commonly grows by rocky streams, but not as close to the water as *S. subalpina*. Common on sedimentary, igneous and metamorphic rocks, especially granite, schist and basalt, but rarely on limestone. Occasionally it grows on walls and tree-trunks, and on sandy heaths and peat. Common associates are *Barbilophozia floerkei*, *Frullania tamarisci*, *Lophozia ventricosa*, *Campylopus introflexus*, *Dicranum scoparium*, *Hypnum cupressiforme*, *Polytrichum* spp., *Racomitrium* spp. and *Cladonia* spp. 0–400 m, absent from higher hills and mountains. GB 427+69*, IR 41+5*.

Paroecious; perianths and sporophytes common. Gemmae frequent.

A Mediterranean-Atlantic species endemic to Europe and nearby territories. Throughout the Mediterranean region including Turkey, N. Africa, Macaronesia and Iberian Peninsula, north to S. Scandinavia, but rare away from the Atlantic coast.

D. G. LONG

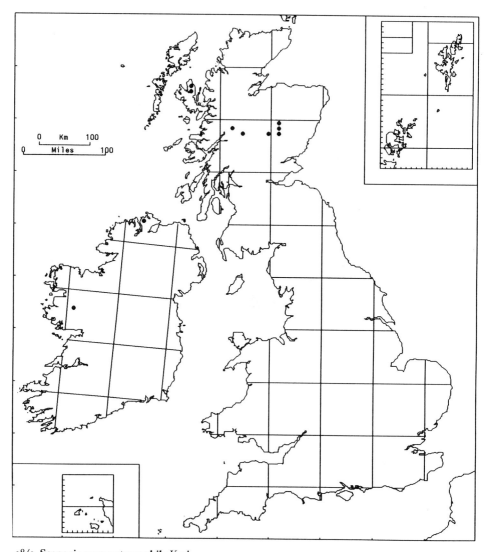

38/2. **Scapania gymnostomophila** Kaal.

A strict calcicole growing in small patches or as scattered stems usually on damp mossy limestone, schist or basalt rock faces, often on shaded N.- or E.-facing slopes, where it grows in earthy rock crevices or mixed with bryophytes such as *Jungermannia subelliptica*, *Preissia quadrata*, *Scapania calcicola*, *Amphidium mougeotii*, *Anoectangium aestivum*, *Gymnostomum aeruginosum*, *Myurella julacea* and *Tortella tortuosa*. In W. Ireland it grows in mossy crevices in low-altitude limestone pavement; in Scotland most localities are in the mountains. 20 m (Co. Mayo) to 950 m (Ben Alder Forest). GB 8, IR 2.

Dioecious and usually sterile; male inflorescences very rare, perianths and sporophytes unknown in the British Isles. Gemmae usually abundant.

Arctic-alpine. European mountains south to Pyrenees, north to Scandinavia, Iceland and Svalbard. Siberia, N. America, Greenland.

Possibly under-recorded.

D. G. Long

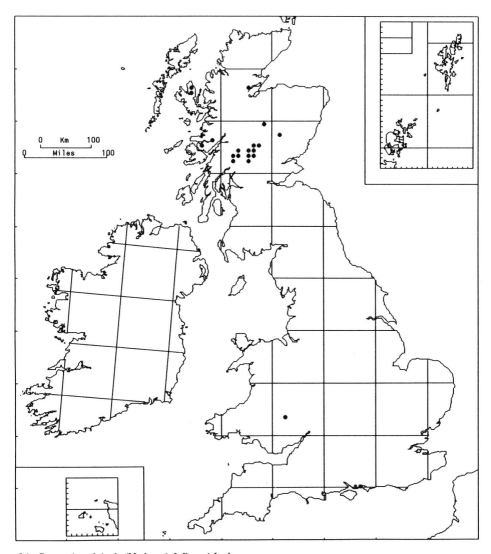

38/3. **Scapania calcicola** (H. Arn. & J. Perss.) Ingham

A diminutive calcicole similar ecologically to *S. gymnostomophila* but generally more abundant, and locally common in the Breadalbane mountains. Habitats include damp shaded mossy cliff ledges, boulders, thin soil overlying rocks, and occasionally eroding or disturbed base-rich soil on mine-waste and near dams. Base-rich schist and soft limestones are most favoured, but it is occasionally found on basalt and calcareous sandstones. It rarely grows directly on the rock but is usually creeping over a thin crust of soil, humus or bryophytes. Associates are many but species such as *Aneura pinguis*, *Blepharostoma trichophyllum*, *Cololejeunea calcarea*, *Jungermannia subelliptica*, *Leiocolea alpestris*, *Scapania aequiloba*, *S. cuspiduligera*, *Preissia quadrata*, *Tritomaria quinquedentata*, *Amphidium mougeotii*, *Anoectangium aestivum*, *Gymnostomum aeruginosum*, *Myurella julacea* and *Tortella tortuosa* are typical. 150 m (Killin) to 1150 m (Ben Lawers). GB 19.

Dioecious; male inflorescences rare, female inflorescences and perianths not found in Britain. Gemmae usually present.

Boreal-montane. Mountains of Europe south to Crimea and Yugoslavia, north to Scandinavia and Iceland. Caucasus, Siberia, eastern N. America.

D. G. Long

187

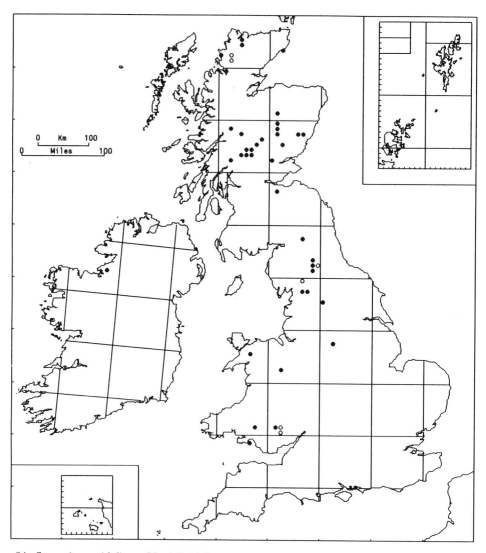

38/4. Scapania cuspiduligera (Nees) K. Müll.

Grows in small whitish-green patches on damp calcareous rocks in shady ravines, and on rocks and detritus on streamsides, montane cliffs and rocky gullies; also on soil and schistose gravel, more rarely on damp calcareous shell-sand dunes. Limestone and basic schist are the usual substrates, but it is also recorded from Old Red Sandstone conglomerate and from Millstone Grit where subject to basic seepage or flooding. It grows with species such as *Blepharostoma trichophyllum, Jungermannia atrovirens, Leiocolea* spp., *Scapania calcicola, Amphidium mougeotii, Blindia acuta* and *Distichium capillaceum*. Sea-level (Hebrides, Sutherland) to 1175 m (Ben Lawers). GB 38+6*, IR 1.

Dioecious; fertile plants uncommon, sporophytes unknown. Gemmae usually present and often abundant.

In the Northern Hemisphere a circumboreal arctic-alpine. Mountains of C. and E. Europe, Pyrenees, Scandinavia, Iceland and Svalbard. Caucasus, Siberia, Mongolia, Japan, N. America, Greenland. Disjunct in Zaire and Colombia.

D. G. LONG

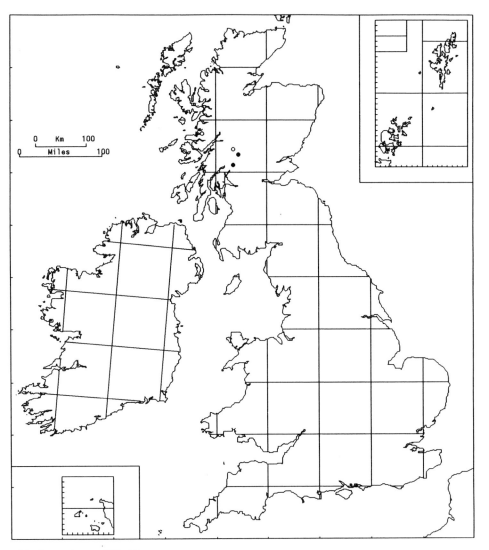

38/5. **Scapania parvifolia** Warnst.

A poorly-known species of bare soil on exposed mountain ridges and summits, with *Cephalozia bicuspidata*, *Diplophyllum albicans*, *Lophozia sudetica*, *Nardia scalaris*, *Scapania scandica* and *Oligotrichum hercynicum*. 600–990 m (An Caisteal). GB 2+2*.

Dioecious; gametangia rare, sporophytes unknown. Gemmae usually present.

Boreal-arctic, circumpolar. Poland, Czechoslovakia, Norway, Sweden, Finland, Svalbard. Siberia, Amur, Japan, Canada.

D. G. Long

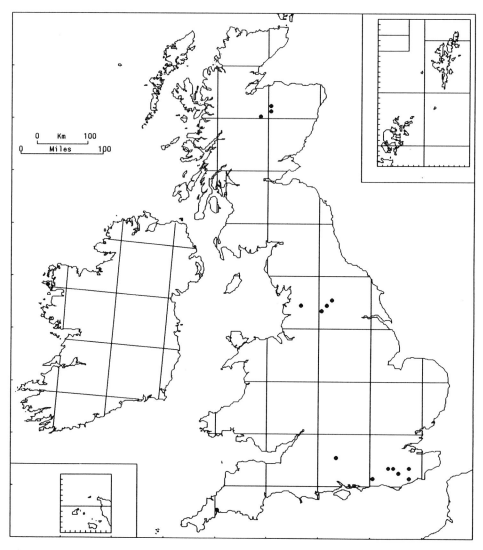

38/6. **Scapania curta** (Mart.) Dum.

A calcifuge pioneer, occurring in similar habitats to *S. scandica*, but more restricted ecologically. It grows on shaded sandstone rock-faces, on sandy ground in woodland rides, old quarries and china-clay works, on peat soil in old pasture, and on detritus by streams. Associated species include *Cephalozia bicuspidata, Diplophyllum albicans, Jungermannia gracillima* and *Scapania irrigua*. Lowland. GB 15, IR 1.

Dioecious; gametangia occasional, sporophytes very rare. Gemmae usually present but not abundant.

Circumboreal. Europe south to Italy and Greece, north to Scandinavia, Iceland, Svalbard. Siberia, N. America, Greenland.

Formerly confused with *S. scandica;* many early literature records are erroneous. Now that it is better understood, it is proving to be quite widespread but certainly much rarer than *S. scandica*.

D. G. LONG

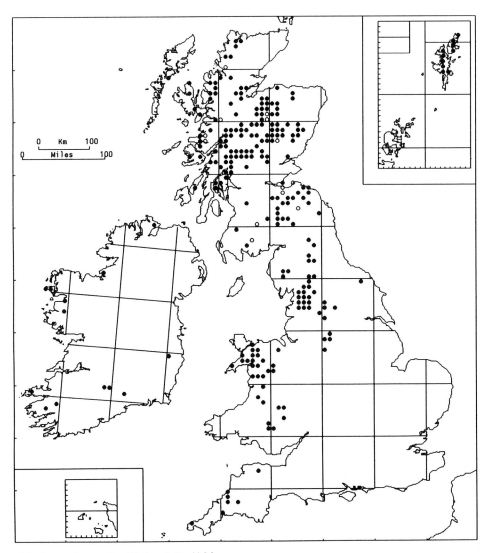

38/7. **Scapania scandica** (H. Arn. & Buch) Macv.

It is usually calcifuge, growing as a pioneer on soil, peat and rock in habitats such as crumbling banks, heathery banks, forestry tracks, footpaths, sand-dunes, moorlands and bogs; occasionally on rotten wood or bark. At higher altitudes it grows on moraine detritus, rocky gullies, streamsides, cliff ledges, stony fell-fields and exposed mountain summits. A wide range of associates is recorded, the most frequent being other calcifuge pioneers such as *Diplophyllum albicans*, *Nardia scalaris*, *Ditrichum heteromallum*, *Hypnum cupressiforme*, *Pogonatum urnigerum* and *Polytrichum juniperinum*, and on exposed summits species such as *Marsupella brevissima* and *Ditrichum zonatum*. 0–1120 m (Creag Meagaidh). GB 225 + 13*, IR 13.

Dioecious; male inflorescences and perianths common, sporophytes rare. Gemmae usually present.

A suboceanic-boreal species widespread in N. Europe and the mountains of C. Europe; Faeroes, Iceland. Azores, N.W. Himalaya, N. America, Greenland.

Formerly considered rare because of confusion with *S. curta*, it is now known to be widespread in the British Isles, though undoubtedly still under-recorded in many districts.

D. G. LONG

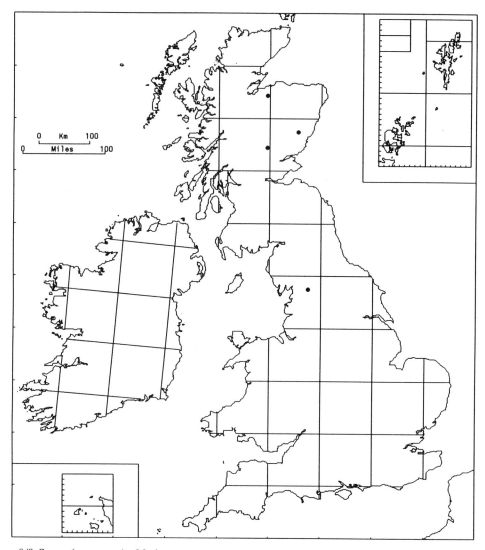

38/8. Scapania praetervisa Meylan

The main habitat is boulders and rock outcrops on river banks, where the substrate is at least mildly basic, for example schist, limestone and sandstone. Recorded associates include *Blepharostoma trichophyllum*, *Diplophyllum albicans*, *Marsupella emarginata*, *Plagiochila porelloides*, *Scapania subalpina* and *Blindia acuta*. 80–135 m. GB 4.

Dioecious; male inflorescences and perianths occasional, sporophytes unknown in the British Isles. Gemmae usually present.

World distribution not fully known owing to confusion with *S. mucronata*. France, Germany, Switzerland, Poland, Czechoslovakia, Sweden and Finland. Novaya Zemlya, Alaska, Greenland.

D. G. LONG

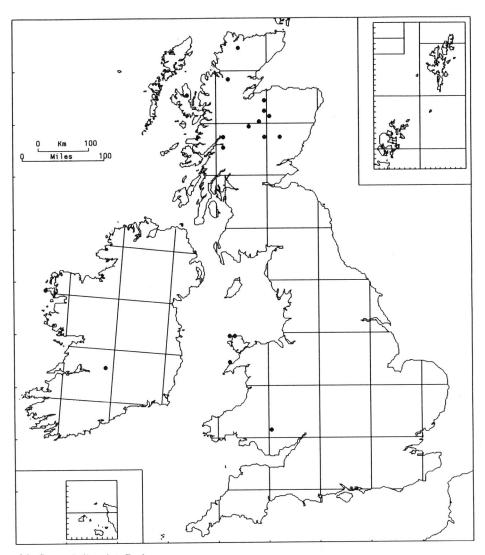

38/9. **Scapania lingulata** Buch

A lowland and subalpine species, it grows as a pioneer on acidic to basic substrates including soil on stream-sides, boulders in ravines, peaty ledges on limestone cliffs, basalt cliff-ledges, and in one site on a dead stem of *Sedum rosea* on basic schist cliffs. Müller (1954–57) describes it as a calcifuge, but this is not the case in Britain. Recorded associates are diverse, including calcifuge pioneers such as *Diplophyllum albicans*, *Marsupella emarginata*, *Nardia scalaris*, *Diphyscium foliosum*, *Ditrichum heteromallum* and *Pogonatum* spp., and on basic substrates *Blepharostoma trichophyllum*, *Leiocolea heterocolpos*, *Amphidium mougeotii* and *Anoectangium aestivum*. Sea-level (Anglesey) to 560 m (Seana Bhraigh). GB 16, IR 2.

Dioecious; gametangia rare, sporophytes unknown. Gemmae usually present.

A boreal species found in N. and E. Europe (rare), Scandinavia, Iceland, U.S.S.R. Eastern N. America, Greenland.

Only recently added to the British flora, its distribution and ecology are still poorly known. The number of records is steadily increasing, so it may prove to be quite widespread.

D. G. LONG

193

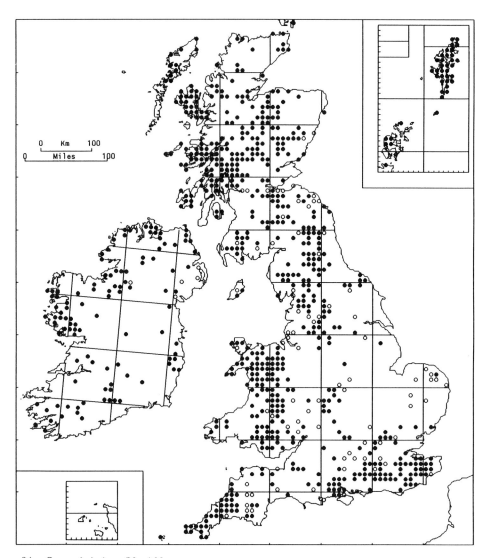

38/10. **Scapania irrigua** (Nees) Nees

Typically a colonist of damp disturbed or compacted soil, clay, mud and gravel on damp footpaths, woodland tracks, old quarries, detritus on streamsides and margins of lakes and reservoirs; also in marshes and boggy pastures. It is occasionally found in coastal dune-slacks. Associated bryophytes include *Pellia* spp., *Riccardia chamedryfolia*, *Calliergon cuspidatum*, *Philonotis fontana* and bulbiliferous *Pohlia* spp. In these habitats it grows generally on acidic and circumneutral substrates. A more robust, erect form grows in damp stony and grassy flushes on mountains, often on base-rich substrates such as limestone and mica-schist. In this habitat it can grow with *Aneura pinguis*, *Drepanocladus revolvens*, *Fissidens adianthoides* and *Sphagnum* spp., especially *S. auriculatum*. 0–975 m (Breadalbane). GB 634+84*, IR 106+6*.

Dioecious; male inflorescences and perianths common, sporophytes rare. Gemmae commonly produced.

A circumboreal and arctic species found throughout Europe, especially in mountainous regions, south to Turkey, north to Scandinavia, Faeroes, Iceland, Svalbard. Caucasus, Siberia, Sakhalin, Japan, N. America, Greenland.

D. G. LONG

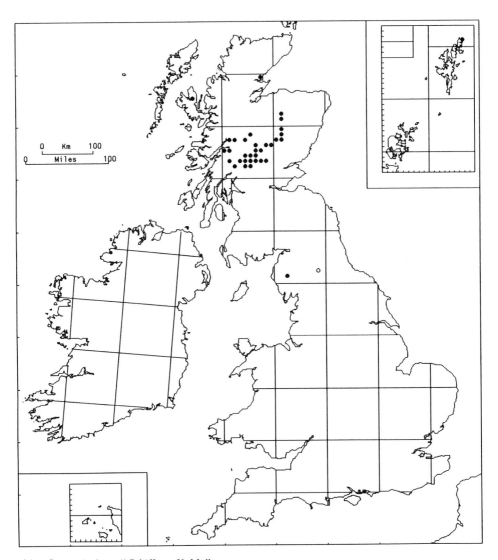

38/11. **Scapania degenii** Schiffn. ex K. Müll.

A strict calcicole favouring substrates derived from basic schist and limestone, rare in Britain except in parts of the Breadalbane mountains. It grows in characteristic green or yellowish tufts in moist mossy turf on damp hillsides, in and beside stony flushes, and on wet rock-ledges and cliffs. Frequent associates are *Selaginella selaginoides*, *Aneura pinguis*, *Blepharostoma trichophyllum*, *Leiocolea bantriensis*, *Scapania aequiloba*, *Bryum pseudotriquetrum*, *Calliergon sarmentosum*, *Campylium stellatum*, *Ctenidium molluscum*, *Ditrichum flexicaule*, *Drepanocladus revolvens*, *Fissidens adianthoides* and *Tortella tortuosa*. Other associates include rare alpine calcicoles such as *Barbilophozia lycopodioides*, *B. quadriloba*, *Meesia uliginosa* and *Oxystegus hibernicus*. 30 m (Shetland) to 950 m (Ben Lawers, Ben Alder Forest), but not often below 400 m. GB 34+1*.

Dioecious; perianths common, male inflorescences and sporophytes unknown. Gemmae usually present.

Arctic-alpine. Alps (rare), Czechoslovakia, Poland, Scandinavia and U.S.S.R. including Siberia. N. America, Greenland.

D. G. Long

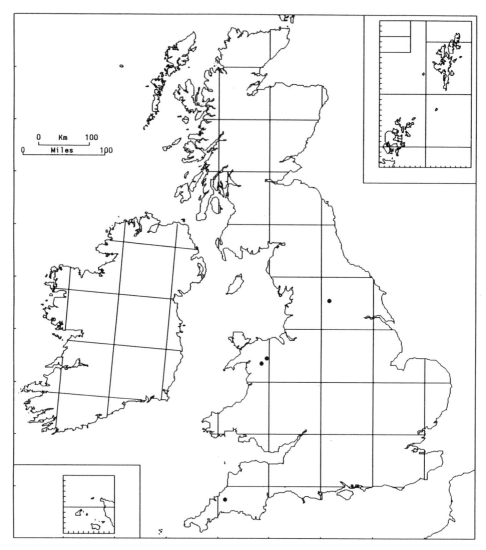

38/12. **Scapania paludicola** Loeske & K. Müll.

Its habitat is lowland and subalpine bogs, acid mires and wet heaths, growing among *Sphagnum* species such as *S. palustre* and *S. papillosum*. 230 m (Cornwall) to 400 m (Merioneth). GB 4.

Dioecious; only male plants have been found in Britain. Gemmae usually present.

A circumboreal-continental species reported from France, Switzerland, Italy, Germany, Czechoslovakia, Poland, Bulgaria, Denmark, Scandinavia, Iceland, Svalbard, U.S.S.R. (including Sakhalin), Japan, N. America, Greenland.

Unaccountably rare in Britain, its eastern distribution in Europe would suggest that it could occur in eastern Scotland.

D. G. Long

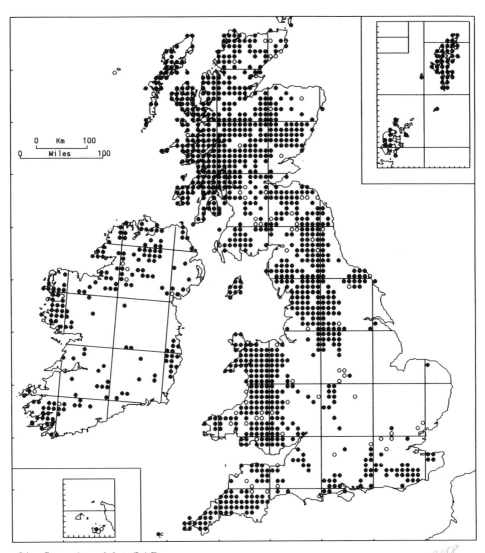

38/13. **Scapania undulata** (L.) Dum.

The most widespread and abundant member of the genus in the British Isles, its occupation of a very diverse range of habitats mirrors its phenotypic plasticity and variability. At lower altitudes it is a plant of constantly or intermittently moist places, especially damp sand, gravel, and boulders by streams, both in open situations and in damp woodlands; also on moist disturbed soil or gravel in old quarries, by woodland tracks and ditches, in seepages on rock-faces and on damp rotten logs in wooded ravines. In the uplands more robust phenotypes are common, particularly in flushes, streams and springs, with *Nardia compressa*, *Scapania uliginosa*, *Calliergon sarmentosum*, *Drepanocladus revolvens*, *Philonotis fontana*, *Pohlia wahlenbergii* var. *glacialis* etc. On wet gravelly ground and streamsides, frequent associates include *Blasia pusilla*, *Marsupella emarginata*, *Scapania subalpina* and *Dichodontium pellucidum*. Although not exclusively calcifuge, it is rarer on basic substrates. 0–1100 m (Ben Nevis). GB 1094+75*, IR 176+9*.

Dioecious; perianths and sporophytes common. Gemmae very common.

A frequent plant through most of Europe, commoner in the mountains and towards the north and west but rare in the east and Mediterranean region. N. Africa, Macaronesia, E. Asia, N. America, Greenland.

D. G. Long

197

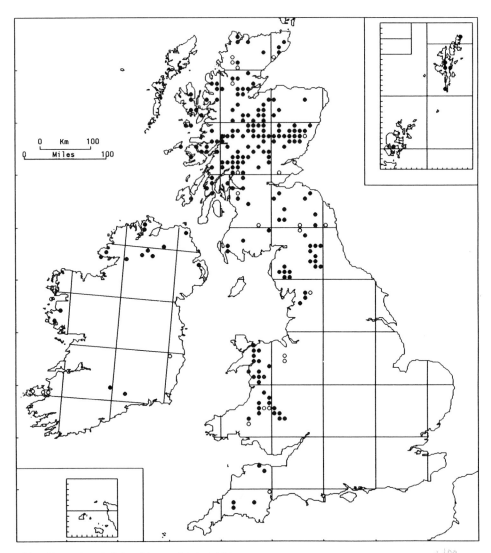

38/14. Scapania subalpina (Nees ex Lindenb.) Dum.

Its most typical home is on damp sandy or gravelly detritus on rocks by moorland and mountain streams, usually quite close to the water. It will also grow on soil or detritus on rocky lake shores, on soil-capped boulders in ravines, and more rarely on mine waste. It is rare on limestone, but is found on most other types of rock including basalt, granite, schists and sandstones. Typical associates are *Blasia pusilla*, *Nardia scalaris*, *Pellia epiphylla*, *Scapania undulata*, *Anomobryum filiforme*, *Brachythecium plumosum*, *Dichodontium pellucidum*, *Pohlia drummondii*, *Polytrichum juniperinum* and *Rhizomnium punctatum*. Generally absent from the lowlands but ranging from 30 m (W. Lothian) to 1180 m (Ben Lawers). GB 201+21*, IR 15+6*.

Dioecious; gametangia common, sporophytes occasional. Gemmae usually present.

A circumboreal montane species. Widespread in the European mountains, south to N. Italy, Pyrenees and north to Scandinavia, Faeroes, Iceland. Madeira, Japan, N. America, Greenland.

D. G. LONG

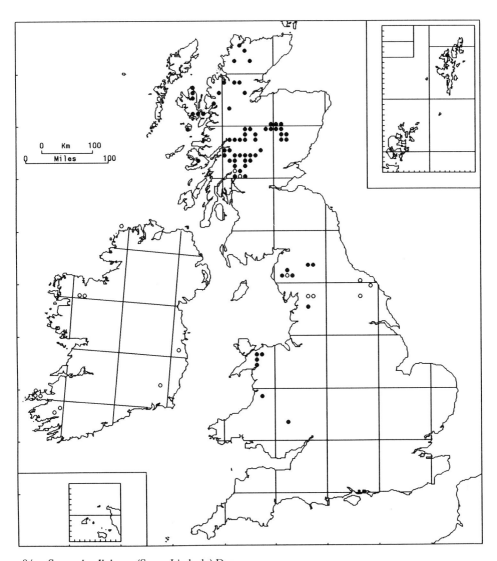

38/15. **Scapania uliginosa** (Sw. ex Lindenb.) Dum.

Grows typically in high-altitude springs and flushes, often submerged or emergent, where it can form extensive conspicuous spongy purplish or reddish-brown patches. Usually in acid to neutral oligotrophic water. It also grows on wet rocks in late-snow areas, in boggy ground, on lake shores, and on constantly irrigated rocks on dripping cliffs and in shaded gullies. Common associates are *Anthelia julacea*, *Harpanthus flotovianus*, *Jungermannia exsertifolia* ssp. *cordifolia*, *Scapania undulata*, *Calliergon sarmentosum*, *Dicranella palustris*, *Philonotis fontana*, *P. seriata* and *Rhizomnium punctatum*. 275 m (Skye) to 1130 m (Ben Nevis). GB 70+9*, IR 10*.

Dioecious; perianths occasional, sporophytes rare. Gemmae unknown in Britain.

Suboceanic arctic-alpine. Mountains of C. and N. Europe, commoner in northern and western areas; east to the Tatra Mts, north to Scandinavia, Faeroes and Iceland. Aleutian Islands, northern N. America, Greenland.

D. G. LONG

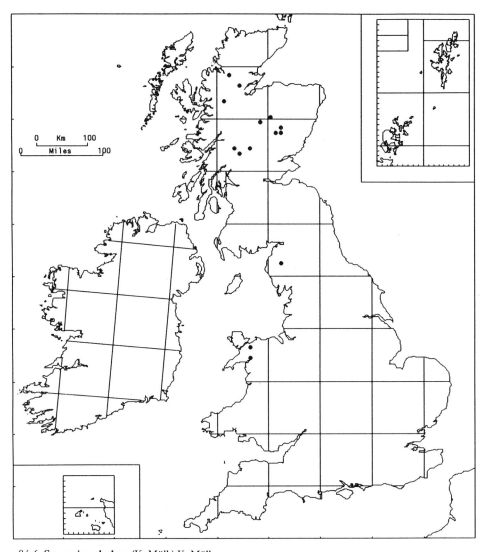

38/16. Scapania paludosa (K. Müll.) K. Müll.

More base-tolerant than *S. uliginosa* but often growing in similar habitats, where it forms robust bright green masses in high-altitude circumneutral or mildly basic bryophyte-dominated springs and flushes; also on wet rocks, especially in rocky gullies on N.- to E.-facing cliffs. Associates include *Dicranella palustris, Philonotis fontana, P. seriata, Pohlia ludwigii* and *Sphagnum palustre*. 330 m (Merioneth), but rarely below 600 m, to 1070 m (Lochnagar). GB 14.

Dioecious; gametangia rare, perianths very rare, sporophytes unknown in Britain. Gemmae absent.

A boreal-alpine suboceanic species with similar range to *S. uliginosa*. Mountains of C. Europe, east to Czechoslovakia, becoming commoner in N.W. Europe; Scandinavia, Faeroes, Iceland. Siberia, Japan, Aleutian Islands, northern N. America, Greenland.

D. G. LONG

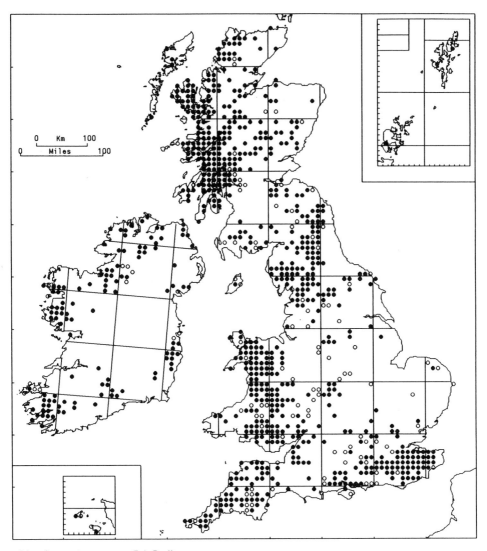

38/17. **Scapania nemorea** (L.) Grolle

In a wide variety of moist or humid habitats, especially lowland broad-leaved woods in valleys and ravines, where it inhabits mossy banks, walls, boulders, rock ledges, logs and stumps, growing on moist ground, rock or rotting wood. Common associates are *Diplophyllum albicans*, *Lophocolea bidentata*, *Lophozia ventricosa*, *Plagiochila porelloides* and *Mnium hornum*. It is also quite common in more open habitats, including marshes, open slopes under *Calluna*, mossy rock-faces and rocky gullies. Acid and neutral substrates are the norm, but it can grow in quite obviously basic habitats, including calcareous rocks and base-rich flushes. 0–600 m (Mweelrea). GB 666+105*, IR 114+13*.

Dioecious; gametangia and sporophytes common. Gemmae usually abundant.

Europe south to Mediterranean and Turkey, north to S. Scandinavia and Baltic States, commonest in the west and absent from the Arctic. Macaronesia, Siberia, E. Asia, Alaska, eastern N. America.

D. G. Long

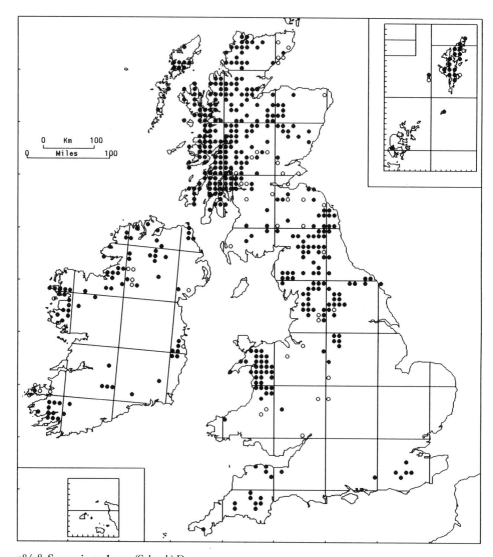

38/18. **Scapania umbrosa** (Schrad.) Dum.

It is most often found growing on decaying wood in damp woodlands and wooded ravines, with *Cephalozia bicuspidata*, *Nowellia curvifolia*, *Riccardia palmata* etc. In subalpine areas it grows quite often on damp bare peat on moorlands and in bogs, with *Calypogeia muelleriana*, *Cephalozia bicuspidata*, *Kurzia* spp., *Lophozia ventricosa*, etc. It is also a typical pioneer of damp rock-faces (especially sandstone) and streamside outcrops, partly buried boulders and slabs, especially in shaded or wooded situations, with *Calypogeia* spp., *Cephalozia bicuspidata*, *Diplophyllum albicans*, *Lepidozia reptans*, *Marsupella emarginata*, *Scapania gracilis*, *Heterocladium heteropterum* and *Mnium hornum*. 0–880 m (Beinn Eighe). GB 401+39*, IR 76+11*.

Dioecious; male inflorescences and perianths common, sporophytes occasional. Gemmae usually abundant.

Widely distributed in mountainous parts of C. and N. Europe, absent from the Arctic, becoming rare southwards to Pyrenees, Azores and Turkey. Japan, N. America.

D. G. LONG

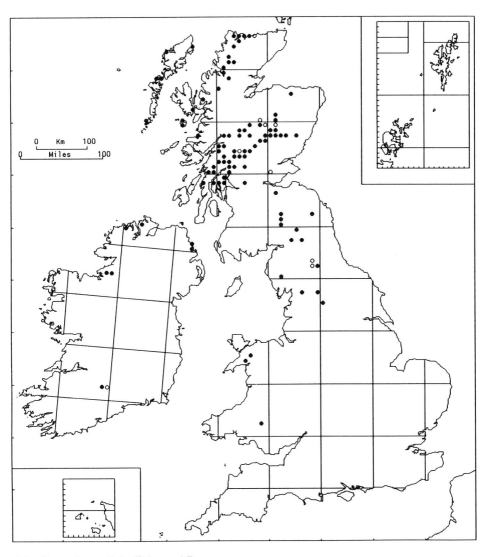

38/19. **Scapania aequiloba** (Schwaegr.) Dum.

It favours well-drained, strongly calcareous substrates, especially schists and limestones, often in rather exposed situations, such as mountain cliff-ledges, boulders, and mossy turfs, but also on rocks in ravines. It is uncommon in moister places such as boulders in streams and flushes. Frequent associates in rocky habitats are *Blepharostoma trichophyllum*, *Cololejeunea calcarea*, *Frullania tamarisci*, *Anoectangium aestivum*, *Barbula ferruginascens*, *Ctenidium molluscum*, *Ditrichum flexicaule*, *Fissidens adianthoides* and *Tortella tortuosa*, whilst in flushes it can grow with *Selaginella selaginoides* and *Scapania degenii*. In the extreme north and west it can grow in damp turf on coastal shell-sand, with *Scapania aspera*, *Cratoneuron filicinum*, *Ctenidium molluscum*, *Ditrichum flexicaule* and *Fissidens cristatus*. 0–1175 m (Ben Lawers). GB 86+9*, IR 7+1*.

Dioecious; perianths occasional, sporophytes unknown in the British Isles. Gemmae usually present.

A European alpine near-endemic, on all calcareous mountain ranges of Europe, south to Spain, Italy, Turkey and Caucasus, north to Scandinavia.

Over-recorded in the past because of confusion with small forms of *S. aspera*. The British records of both species have been revised by Long (1978).

D. G. LONG

203

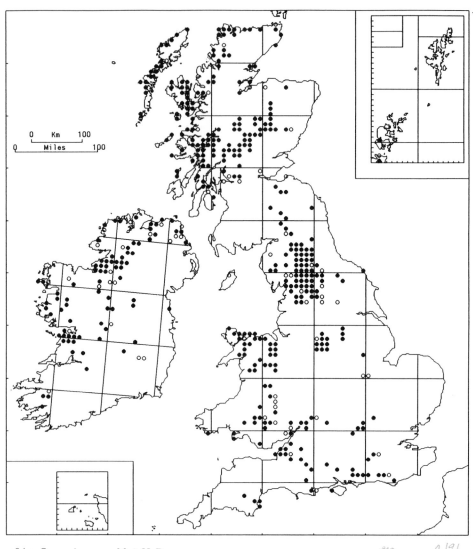

38/20. **Scapania aspera** M. & H. Bern.

A more southern, less montane taxon than *S. aequiloba*, but otherwise quite similar in its ecological requirements. It is abundant in the limestone regions of N. England; elsewhere it is only locally common on limestones, chalk, basic schists, basic igneous and basic sandstone rocks and soil, in chalk or limestone turf, and not infrequently on calcareous sand-dunes. On rocks, associates include many common calcicoles, in particular *Ctenidium molluscum*, *Distichium capillaceum*, *Ditrichum flexicaule* and *Tortella tortuosa*; in basic turf they include *Selaginella selaginoides*, *Breutelia chrysocoma*, *Ctenidium molluscum*, *Pseudoscleropodium purum* and *Rhytidiadelphus triquetrus*; and on sand-dunes *Barbula fallax*, *Ctenidium molluscum* and *Ditrichum flexicaule*. 0–1040 m (Ben Lawers). GB 315+47*, 77+21*.

Dioecious; perianths and male inflorescences common, sporophytes rare. Gemmae usually abundant.

European near-endemic with distribution similar to that of *S. aequiloba*, but rarer in the north and commoner in the south, to Sicily, Yugoslavia, Greece, Turkey.

At lower altitudes the plants are usually robust and distinctive, but at higher altitudes diminutive forms can be deceptively similar to *S. aequiloba*.

D. G. Long

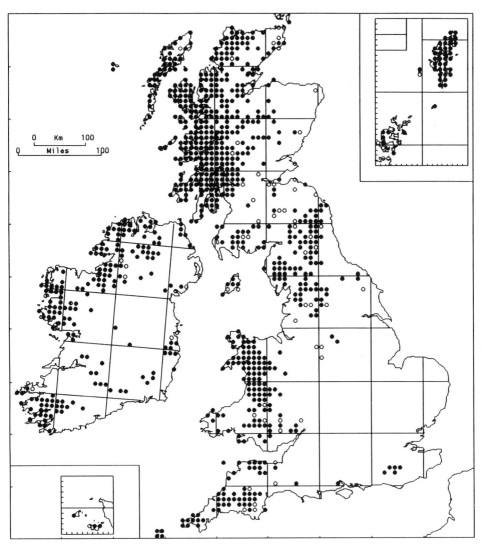

38/21. **Scapania gracilis** Lindb.

A conspicuous hepatic abundant over much of N. and W. Britain and Ireland, in a wide range of habitats. It is particularly common in rocky places, especially on acidic but sometimes on mildly basic substrates, both in woodland and in more open situations. Habitats include coastal rocky banks and gullies, woods, exposed rocky hillsides, cliffs, block-screes and rocky lake-margins. In woodland, it often grows in extensive sheets on boulders, logs, stumps and living tree-trunks, with associates such as *Hymenophyllum wilsonii*, *Frullania tamarisci*, *Plagiochila punctata* and *P. spinulosa*. It is usually a conspicuous component of oceanic-montane bryophyte mats, both in dwarf-shrub heaths and in sheltered block-screes, with *Anastrepta orcadensis*, *Bazzania tricrenata*, *Herbertus aduncus*, *Pleurozia purpurea* etc., and with rarities such as *Anastrophyllum donnianum*, *Mastigophora woodsii* and *Plagiochila carringtonii*. 0–730 m (Skye). GB 682+58*, IR 179+10*.

Dioecious; plants often fertile, perianths and sporophytes frequent. Gemmae usually abundant.

An oceanic European species found throughout the coastal regions and adjacent mountainous parts of W. Europe, from Portugal and Spain to Norway and Sweden; also in Italy. N. Africa, Macaronesia.

D. G. LONG

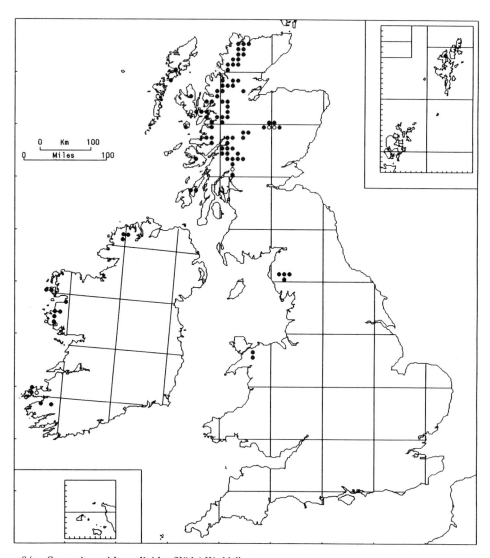

38/22. Scapania ornithopodioides (With.) Waddell

A distinctive species growing as scattered stems amongst other bryophytes or as pure dark reddish-brown patches at moderate to high altitudes in mossy block-screes, on rocky well-drained slopes, on ledges on or at base of low cliffs, and in dwarf-shrub heaths. Most commonly found on acid substrates on open slopes or in mountain corries with a N. to E. aspect, where shade and humidity are maintained. In the Cairngorms it grows in mossy block-screes in late-snow areas. All sites enjoy a cool wet climate. Characteristic associates include a range of other oceanic-montane hepatics such as *Anastrepta orcadensis*, *Anastrophyllum donnianum*, *Bazzania pearsonii*, *Mastigophora woodsii* and *Pleurozia purpurea*, and commoner species such as *Bazzania tricrenata*, *Diplophyllum albicans*, *Mylia taylorii*, *Scapania gracilis* and *Racomitrium lanuginosum*. 230 m (Mull) to 965 m (Beinn Ime). GB 78+5*, IR 14+2*.

Dioecious; usually sterile, perianths very rare. Gemmae occasional.

Disjunct in cool-temperate regions of the Northern Hemisphere and on tropical islands. Norway, Faeroes. Himalaya, W. China, Japan, Taiwan, Philippines, Hawaii. Absent from N. America.

D. G. Long

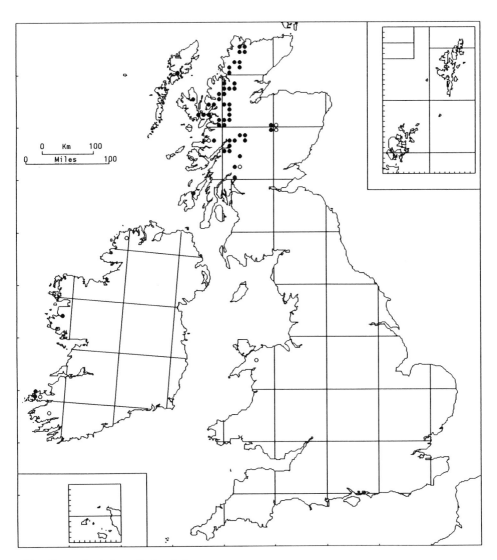

38/23. Scapania nimbosa Tayl. ex Lehm.

It grows as loose coppery-red tufts, or mixed with other bryophtes in turf on open well-drained rocky slopes and in mossy block-screes, usually in sheltered situations or in mountain corries facing north to east, more rarely on acid cliff-ledges. Its requirements are similar to those of *S. ornithopodioides*, but it is a rarer species, usually present in smaller quantities. Associates include *Anastrepta orcadensis, Anastrophyllum donnianum, A. joergensenii, Bazzania pearsonii, B. tricrenata, Mylia taylorii, Plagiochila carringtonii, Scapania ornithopodioides, Racomitrium lanuginosum* and *Rhytidiadelphus loreus*. All localities have a cool wet climate, and often enjoy late snow-lie, especially in the Cairngorms. It is a typical member of the mixed northern hepatic mat of N.W. Scotland. 410 m (Applecross) to 980 m (Ben Lui). GB 48+5*, IR 3+4*.

Sterile; gametangia and sporophytes unknown. Gemmae very rare.

A relict disjunct species of cool-temperate high-rainfall areas; outside the British Isles known only from S.W. Norway, E. Himalaya (Nepal, Sikkim) and W. China (Yunnan).

D. G. LONG

207

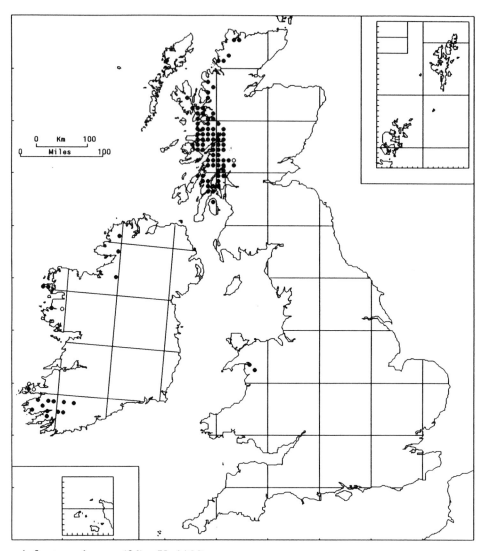

39/1. **Leptoscyphus cuneifolius** (Hook.) Mitt.

This species occurs most frequently as an epiphyte on trunks of trees, especially birch but occasionally oak or willow, and more rarely rowan, in natural or semi-natural birch, birch-hazel, or birch-oak woods on steep blocky slopes and in and along sheltered ravines. It is commonly associated with *Douinia ovata, Frullania tamarisci, F. teneriffae, Plagiochila exigua, P. punctata, P. spinulosa* and *Scapania gracilis*. It also occurs on stems of tall shaggy heather on steep N.- or E.-facing block-strewn slopes, and, more rarely, on damp sheltered but not deeply shaded rocks and cliffs. In Scotland and Wales it occurs from 0–330 m, but in S.W. Ireland it ascends to at least 760 m. GB 81+2*, IR 18+4*.

Dioecious, sex organs very rare; female plants have been found in Killarney (Paton, 1967b). Vegetative propagation is by shoot fragmentation and shedding of leaves.

A widespread Atlantic species, recorded from W. Norway, Azores, Madeira, Tristan da Cunha and N. America.

Because it is easily overlooked, it is probably commoner in W. Scotland and Ireland than the map suggests.

H. J. B. Birks

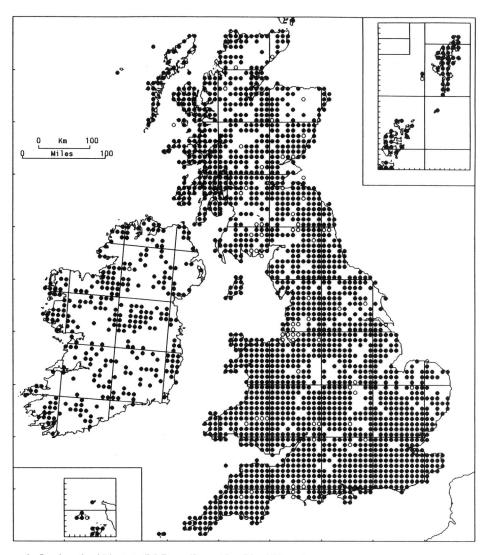

40/1. **Lophocolea bidentata** (L.) Dum. (*L. cuspidata* (Nees) Limpr.)

The species is found in grass, especially in damp situations, on shaded ground, on rotting wood, on the roots and boles of trees and on mossy rocks, walls and boulders. Var. *rivularis* (Raddi) Warnst., the less common variety, is most frequently on damp ground, var. *bidentata* on dead or living wood, but they overlap ecologically. To 850 m on Creag Meagaidh. GB 2025+76*, IR 307+3*.

Var. *bidentata* is autoecious, very often with capsules, ripe in spring and early summer. Var. *rivularis* is dioecious and capsules are rare. Gemmae have been reported in both varieties but are very rare.

Circumboreal. The species is recorded in Europe from mid-Scandinavia south to the mountains of the Mediterranean region. Also in tropical Africa.

The two varieties have been mapped together because of the difficulty in distinguishing them (see Steel, 1978). Recording of the varieties has been unreliable both in Britain and on the Continent.

A. C. CRUNDWELL

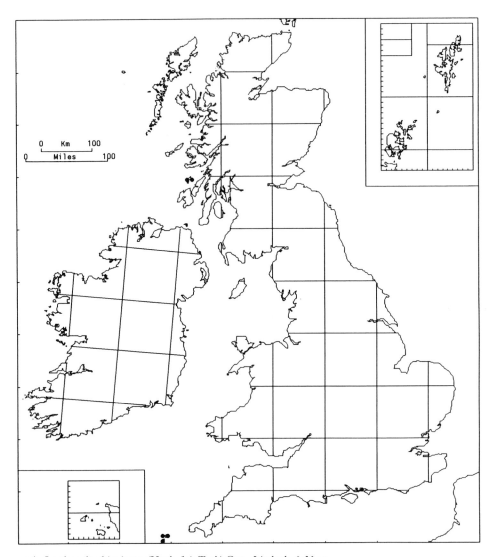

40/2. **Lophocolea bispinosa** (Hook. f. & Tayl.) Gott., Lindenb. & Nees

On thin layers of soil on rocks, walls and stones, on peaty and sandy banks and tracks, in moist sandy depressions, occasionally on tree-trunks. Sometimes associated with *L. semiteres*. Shade-tolerant. Lowland. GB 5.

Dioecious. Sporophytes occasional, on Scilly maturing in spring. Dispersed by caducous branches, gemmae unknown.

An introduced species widespread in Australasia. First gathered in Britain in 1962 from the Abbey Gardens on Tresco, Scilly (Paton, 1974b). It is now abundant there and has spread to all four of the other inhabited islands. The Scottish record is due to E. C. Wallace who found it in 1978 on the Isle of Colonsay, in a garden and in a ravine half a mile away (Wallace, 1979). Probably introduced independently in both regions with the planting of exotic trees.

A. C. CRUNDWELL

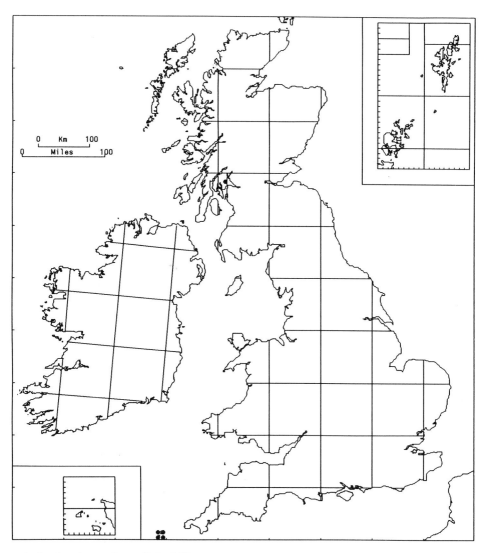

40/3. **Lophocolea semiteres** (Lehm.) Mitt.

Forming mats around the bases of trees, on rotting logs and shaded peaty banks, on paths and in depressions in heathland and on fixed non-basic dunes. Lowland. GB 5.

Dioecious and, rarely, gemmiferous. On Scilly female plants are very rare. In Scotland plants of both sexes are frequent and sporophytes occur.

Introduced. Native in S. America, S. Africa and Australasia. First British record from Tresco in 1955 (Paton, 1965b). It is now plentiful there and on neighbouring islands. Abundant in Benmore Gardens, where first found in 1972 (Long, 1982), but not yet known outside them. Probably introduced independently in both regions with the planting of exotic trees.

A. C. CRUNDWELL

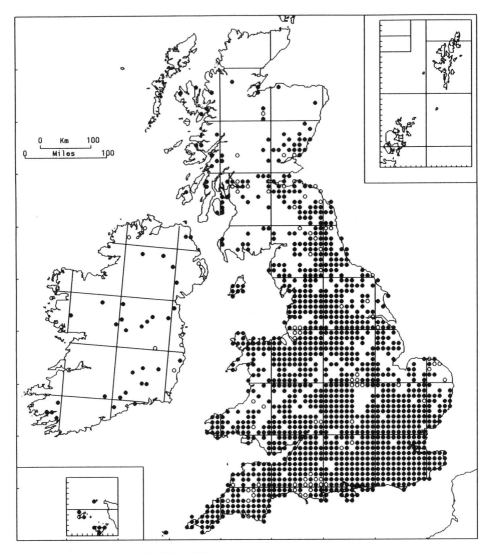

40/4. Lophocolea heterophylla (Schrad.) Dum.

On rotting wood, on tree-boles and trunks, occasionally on rocks, sometimes on soil in shade. More tolerant of atmospheric pollution than most Jungermanniales. In S. and E. England the commonest liverwort species. Lowland. GB 1279+77*, IR 32+8*.

Paroecious, rarely autoecious; capsules abundant, ripe in spring and early summer. Gemmae occasional. Circumboreal. In Europe from mid-Scandinavia southwards, less frequent in the Mediterranean region.

A. C. CRUNDWELL

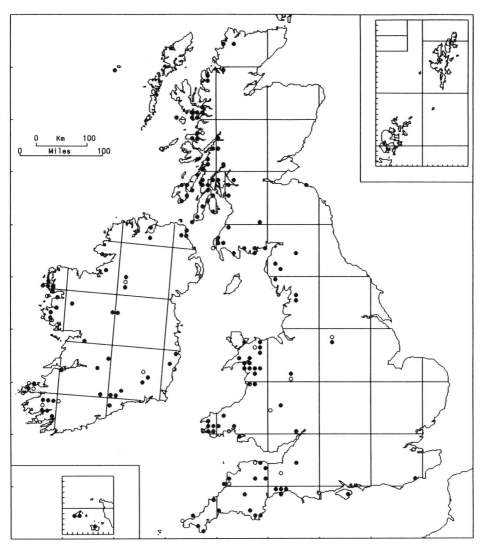

40/5. Lophocolea fragrans (Moris & De Not.) Gott., Lindenb. & Nees

A species of damp, shaded rocks in woods and ravines, sometimes also on trees and rotting branches. Often in dripping caves by the sea, though probably not especially salt-tolerant. Lowland. GB 111+15*, IR 42+12*.

Autoecious. Capsules occasional, maturing in the spring. Gemmae reported by Müller ('stets vorhanden') but not apparently seen by anyone else.

Oceanic species extending from S.W. Norway and N. France to Corsica, Morocco and the Canaries.

A. C. CRUNDWELL

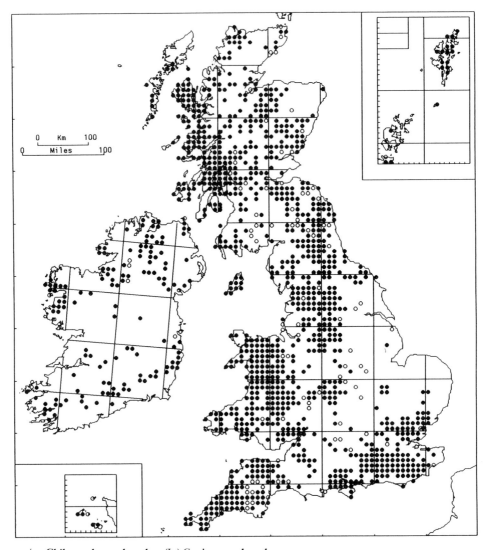

41/1a. **Chiloscyphus polyanthos** (L.) Corda var. **polyanthos**

Occurs most frequently on rocks and boulders in and by streams and rivers, sometimes growing submerged, where it is associated with species such as *Scapania undulata* and *Rhynchostegium riparioides*. It is also found in a similar range of habitats to var. *pallescens*, including marshes, flushes, pool and lake margins, stream-banks, damp soil in woodland (where it can tolerate considerable shade), rotting wood and wet cliff-ledges. Localities vary from mildly basic to moderately acidic. 0–800 m (Glen Lyon). GB 1024+87*, IR 135+10*.

Autoecious; capsules occasional, ripe spring.

Circumboreal.

Sterile *Chiloscyphus* can be very difficult or sometimes impossible to name, and there are likely to be errors in the maps of both species. However, the overall distribution patterns are almost certainly correct.

M. M. YEO

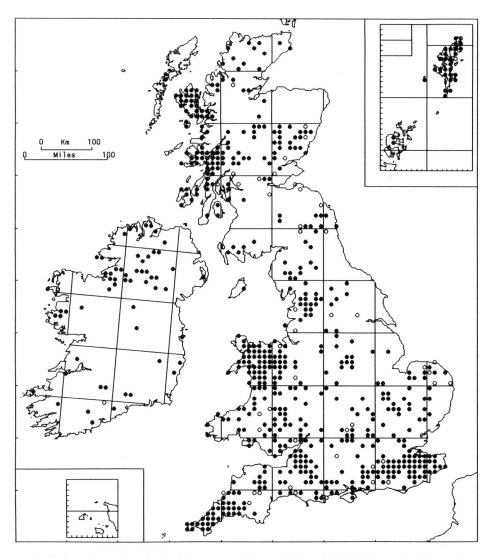

41/1b. Chiloscyphus polyanthos (L.) Corda var. **pallescens** (Ehrh. ex Hofffm.) Hartm. (*C. pallescens* (Ehrh. ex Hoffm.) Dum.)

The chief habitats are marshes, fens, flushes and upland springs, where it grows either among higher plants such as *Carex* and *Juncus* species, or in bryophyte-dominated vegetation. It also grows in alder and willow carr (where it is frequently associated with *Calliergon cordifolium*), on damp soil and rotting logs in woodland, on soil and rocks by lakes and rivers, in damp grassland and on cliff ledges. It is most frequent in situations that are at least mildly basic. 0–850 m (Cam Creag). GB 563+49*, IR 57+2*.

Autoecious; capsules are apparently rare, ripe spring.

Circumboreal. Widespread in Europe but rare in the extreme south.

M. M. YEO

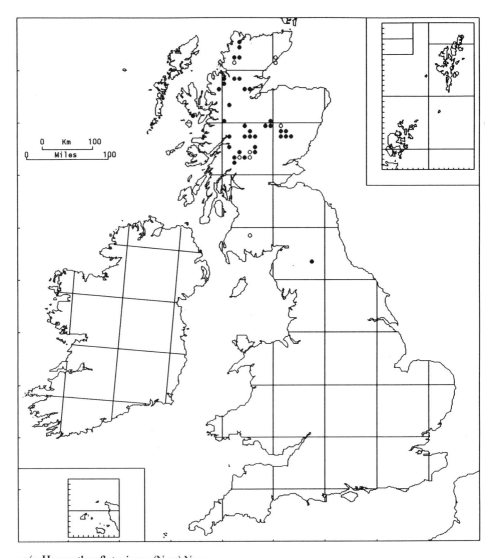

42/1. **Harpanthus flotovianus** (Nees) Nees

A species of mildly basic flushes and springs associated with *Saxifraga stellaris*, *Chiloscyphus* spp., *Scapania undulata*, *Bryum weigelii*, *Dicranella palustris*, *Philonotis fontana* and, more rarely, *Tritomaria polita* and *Oncophorus virens*. It also occurs at the side of streams, on wet banks, and in damp turf. It is rarely present in any abundance and commonly occurs as scattered small pure patches. It extends from near sea-level in Shetland to nearly 1000 m in the Breadalbane mountains and the Cairngorms. It is most frequent above 450 m. GB 36+11*.

Dioecious; female inflorescence common, male uncommon. Capsules rare.

Occurs widely in montane regions of C. Europe and northern areas of Europe, extending from Poland and Yugoslavia to northern Fennoscandia. Siberia, E. Asia, N. America.

H. J. B. Birks

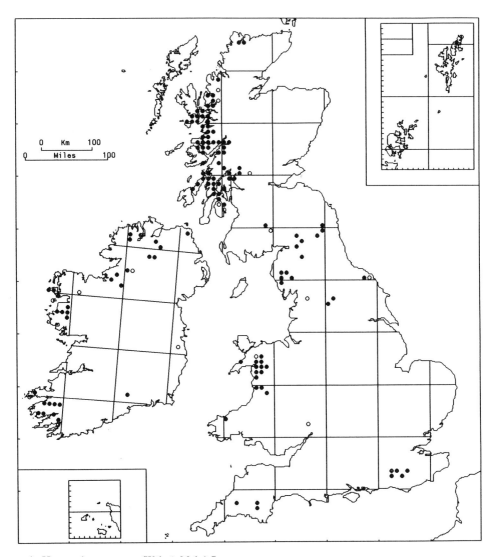

42/2. Harpanthus scutatus (Web. & Mohr) Spruce

It usually occurs in small quantity in a range of shaded habitats, commonly N.- or E.-facing. These include damp peat- or humus-banks near the sea with *Mylia taylorii, Calypogeia neesiana, Lepidozia cupressina* and *Saccogyna viticulosa;* on dry moss-covered acid boulders in deciduous woods on steep, block-strewn slopes; on rather bare steep blocks in woods that are intermittently irrigated by water where it commonly occurs with *Scapania umbrosa, Heterocladium heteropterum* and *Hypnum callichroum;* on shaded sandstone rocks and blocks that are continually moist with *Calypogeia integristipula, Kurzia sylvatica* and *Tritomaria exsecta;* and on decaying logs in deciduous woods with *Jamesoniella autumnalis, Lepidozia reptans* and *Tritomaria* spp. It favours low-lying areas, but ascends to 350 m in S.W. Ireland. GB 101+12*, IR 29+7*.

Dioecious, fertile.

It occurs widely throughout Europe and does not have any strongly western distribution within Europe, extending from Romania, Bulgaria and Yugoslavia northwards to Finland, Iceland, Russia. It also occurs in E. Asia and N. America.

H. J. B. BIRKS

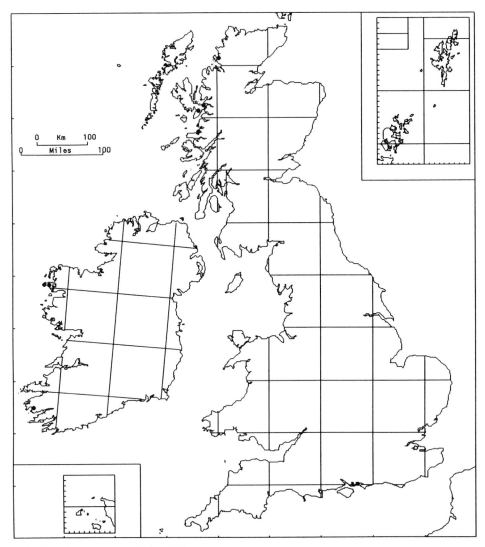

43/1. Geocalyx graveolens (Schrad.) Nees

It is confined to a few shaded, sheltered N.-facing moist humus-banks and rocks near the sea. It occurs mixed with other bryophytes on damp, humus-rich soil and is commonly associated with *Dryopteris aemula, Harpanthus scutatus, Lepidozia cupressina, Mylia taylorii, Saccogyna viticulosa, Scapania gracilis* and *Plagiothecium undulatum.* GB 4, IR 3.

Monoecious; sporophytes occasional.

Widely, but sparsely, distributed throughout Europe (Spain, France, Switzerland, Italy, Greece, Romania, Austria, Czechoslovakia, Poland, U.S.S.R., Netherlands, Denmark, Norway, Sweden, Finland). N. America and Siberia.

Its extreme rarity in the British Isles is curious considering the widespread nature of its habitat in the west. Its extreme western distribution is also inexplicable in view of its mainland European distribution. It occurs in small quantity in its known localities in Co. Kerry and on Skye, is more frequent in its mainland Scottish stations, and is reported to be locally common in its newly-discovered localities on Achill Island.

H. J. B. BIRKS

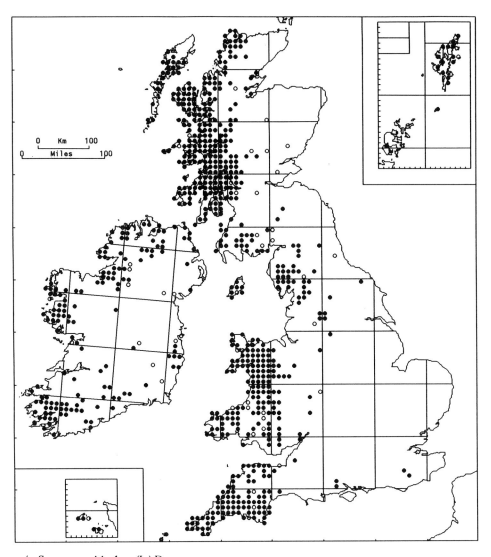

44/1. Saccogyna viticulosa (L.) Dum.

Locally abundant on shaded, slightly moist boulders in deciduous woods (oak, birch, hazel, ash), on damp shaded ledges, banks, and rocks in wooded low-lying ravines and gullies, on and amongst boulders in sheltered block-litters, usually N.- or E.-facing, and on steep, sloping rocks where there is some water seepage and some shade. It also occurs amongst other bryophytes on shaded humus-banks by the sea, on rocky slopes near the sea, on sheltered sea-cliffs, and, more rarely, in damp crevices in granite tors, in hepatic-rich dwarf-shrub heath on steep N.- or E.-facing block-strewn slopes, and on shaded peaty banks. Indifferent to rock type, often most luxuriant in slightly basic situations but rare on limestone. Mainly in the lowlands (0–400 m); in Co. Kerry it ascends to at least 800 m. GB 574+36*, IR 146+12*.

Dioecious; fruit very rare.

Mediterranean and W. Europe, extending from Corsica, Sardinia, and Italy, north to S.W. Norway, Faeroes. Azores, Canaries, Madeira.

H. J. B. Birks

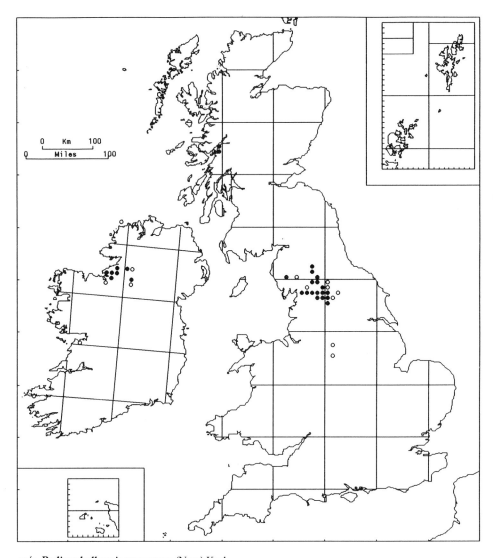

45/1. **Pedinophyllum interruptum** (Nees) Kaal.

On sheltered moist Carboniferous limestone rock ledges and ravine walls, less often on boulders. In Scotland it occurs rarely on metamorphic limestone; in one English site it grows on Millstone Grit where receiving basic floodwater. Mainly lowland, but submontane in the Craven Pennines (to 500 m) and in the Benbulbin range. GB 19+10*, IR 7+6*.

Autoecious; perianths rare, sporophytes very rare; vegetative propagules not known.

Rare in Europe, from Spain, Corsica and Greece north to S. Sweden and W. Russia. Caucasus, Algeria, eastern N. America.

Over-recorded in Britain and elsewhere in the past because of confusion with *Plagiochila porelloides*, forms of which with entire or sparsely dentate leaves often occur in the same habitat. For a map of its European distribution based on expertly-determined material, see Grolle (1969).

A. J. E. SMITH

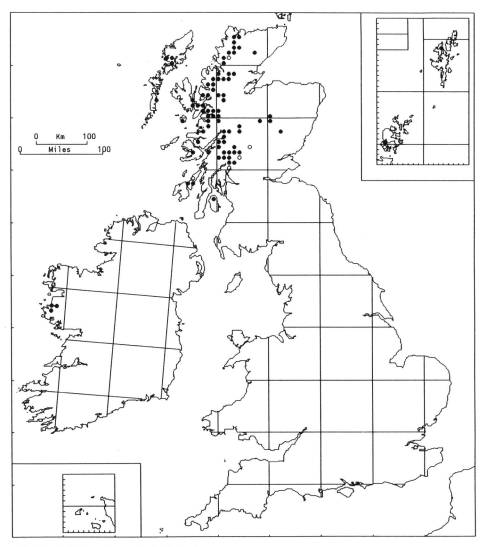

46/1. Plagiochila carringtonii (Balf.) Grolle

A montane species of oligotrophic dwarf-shrub communities on well drained slopes; also occurring among block screes, on steep rock ledges, in acid grassland, and on rocky banks in the upper parts of steep, rocky woodland. Most localities have a N. to E. aspect, and all have a cool and extremely humid climate, except in the Cairngorms, where it is confined to areas of late snow-lie. 150–940 m (Conival). GB 75+6*, IR 3.

Dioecious; only male plants known in Europe.

In Europe known only from the British Isles and the Faeroes. Disjunct in Nepal.

The Nepalese population is female and has completely entire leaves; it has been distinguished as ssp. *lobuchensis* by Grolle (1964). Nevertheless, *P. carringtonii* must clearly rank as one of the most notable disjunctive large liverworts of cool regions with high rainfall. Its habitat and distribution are very similar to those of *Scapania nimbosa*.

B. Averis

221

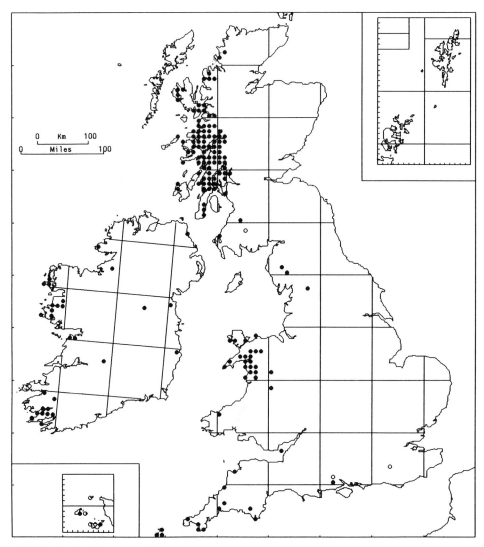

46/2. **Plagiochila killarniensis** Pears.

Locally frequent in deciduous woodland, where it grows usually in pure tufts or patches on basic to mildly acidic rock outcrops, and on the trunks and branches of trees, especially ash and hazel. It is also found on sheltered rock-faces in more open situations. In the northern part of its range it favours a S.W. to S.E. aspect. All localities have a mild and humid climate. Mostly lowland, it ascends to 430 m in S.W. Ireland. GB 135+5*, IR 29+1*.

Dioecious; both male and female plants occur in the British Isles, and have been found intermixed; sporophytes unknown.

Atlantic seaboard of Europe from Scotland south to Spain and Portugal; very rare further east: in Belgium (extinct), Luxemburg (one locality) and N. Italy (no recent record). Azores, Madeira, Canaries.

For a map of the European distribution, refer to Grolle & Schumacker (1982). Paton (1977b) gives more detailed information on its habitat.

B. Averis

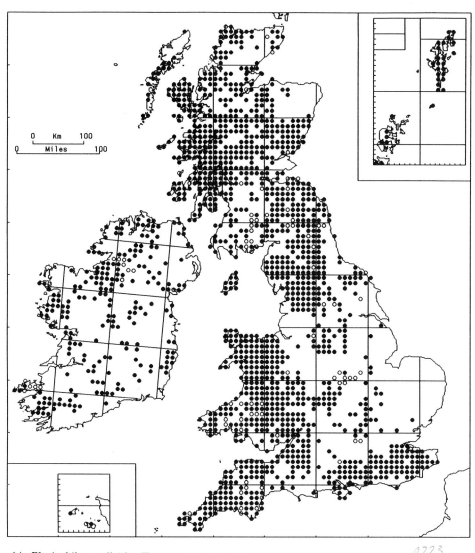

46/3. **Plagiochila porelloides** (Torrey ex Nees) Lindenb.

p223

On basic to mildly acidic rock-faces and rocky banks in woodland, on stone walls, by shaded streams, and on rocks and banks in the uplands. It also occurs on soil-banks in woods, hedgerows and limestone grassland, in sand-dunes, and occasionally on trees. Very variable in size and shape, it can form pure tufts, or grow mixed with other bryophytes. 0–1200 m (Ben Lawers). GB 1302+72*, IR 219+13*.

Dioecious; male inflorescences and perianths frequent, sporophytes rather rare.

Circumboreal. Throughout Europe including the Arctic, becoming submontane in the south.

B. AVERIS

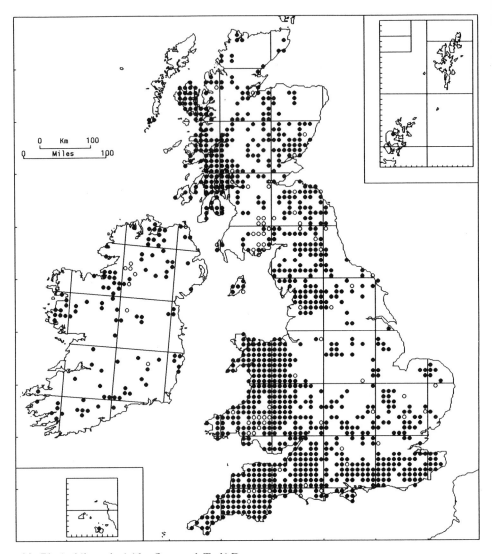

46/4. **Plagiochila asplenioides** (L. emend. Tayl.) Dum.

Occurring in pure patches or as scattered shoots mixed with other large bryophytes on shaded soil or humus banks in lowland woods, wooded ravines and hedgerows. It also occurs in more open situations in species-rich neutral or calcareous turf on N.- or E.-facing slopes. 0–610 m (Schiehallion). GB 1018+69*, IR 108+8*.

Dioecious; male inflorescences and perianths rare, capsules very rare.

Most of Europe but montane in the south and absent from the Arctic. W. Asia.

B. AVERIS

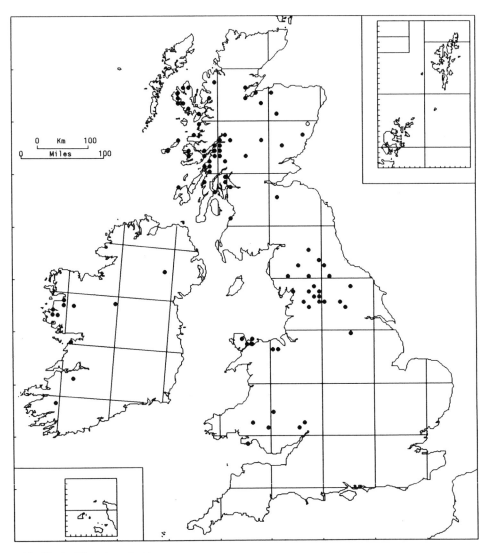

46/5. **Plagiochila britannica** Paton

A calcicole, occurring in pure patches or mixed with other large bryophytes, especially *P. porelloides*, on old shady walls and rocky banks in deciduous woodland and wooded ravines, and, less commonly, on steep rock-faces in open but sheltered situations, usually with a N. to E. aspect. It also grows on tree-bases and detritus by calcareous streams, and in loose bryophyte mats on the ground in base-rich woodland. Mainly lowland, to 360 m near Tomintoul. GB 83+1*, IR 11.

Dioecious; male inflorescences and perianths frequent, sporophytes very rare.

Although this recently recognized species is so far known only from the British Isles, where it has been overlooked as *P. porelloides* or *P. asplenioides*, it is most unlikely to be a British endemic.

B. AVERIS

225

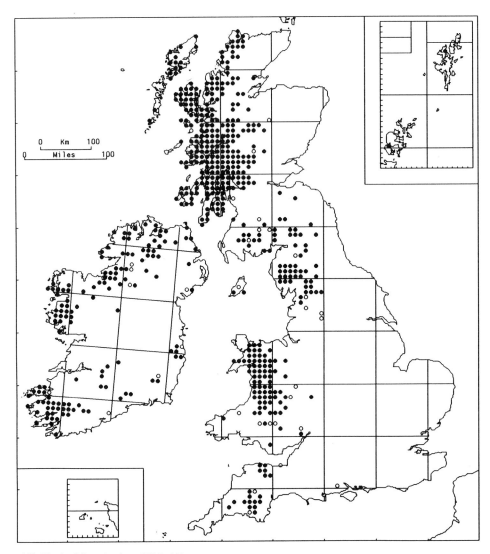

46/6. **Plagiochila spinulosa** (Dicks.) Dum.

A characteristic species of western deciduous woodland, where it can form large tufts or cushions on boulders, rock outcrops, humus banks and tree-trunks, often in association with *Bazzania trilobata* and *Scapania gracilis*. It also occurs on sheltered rock-faces and banks on open rocky slopes in the uplands, ascending to 730 m (Skye). GB 421+25*, IR 133+9*.

Dioecious; female in the British Isles, sporophytes unknown. Presumably spreads by deciduous leaves.

An Atlantic species, very rare in Europe except in the British Isles; known from C. France, Brittany, Normandy, Belgium, Luxemburg, S.W. Norway, Faeroes. Azores, Canaries, Madeira.

For a map of its world distribution, refer to Grolle & Schumacker (1982).

B. Averis

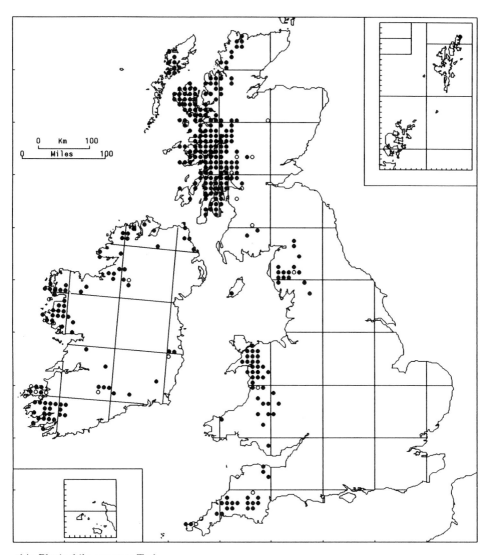

46/7. Plagiochila punctata Tayl.

Most frequent in deciduous woodland, growing usually as pure tufts or cushions on the trunks and branches of trees, especially birch, oak and alder. It is also frequent on rock-faces in woodland and, less commonly, in more open but sheltered rocky situations. 0–580 m in Scotland (Skye) and to 650 m in S.W. Ireland (Co. Kerry). GB 263+13*, IR 83+10*.

Dioecious. Male inflorescences known only from Killarney; perianths occasional; sporophytes unknown. It presumably spreads clonally by its deciduous leaves.

Atlantic seaboard of Europe from S.W. Norway (but not Faeroes) south to Spain and Portugal and east to the Ardennes. Azores, Canaries, Madeira.

B. AVERIS

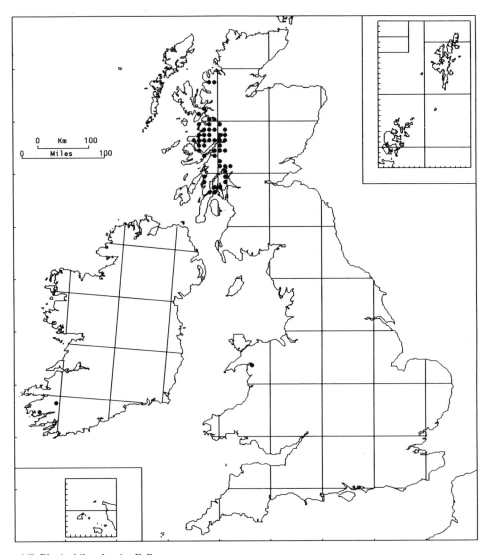

46/8. **Plagiochila atlantica** F. Rose

A species of humid deciduous woodland, occurring mostly on acidic boulders, rock outcrops and on the trunks of oak and birch trees. Occasionally it grows on other trees and on humus banks, rarely on logs. Usually in pure patches, it is often associated with other large bryophytes such as *Plagiochila spinulosa* and *Scapania gracilis*. It favours a S. to E. aspect. 0–200 m (N. Wales). GB 48, IR 2+1*.

Dioecious; only female plants are known, perianths frequent.

Outside the British Isles known only from a single site in N.W. France.

B. AVERIS

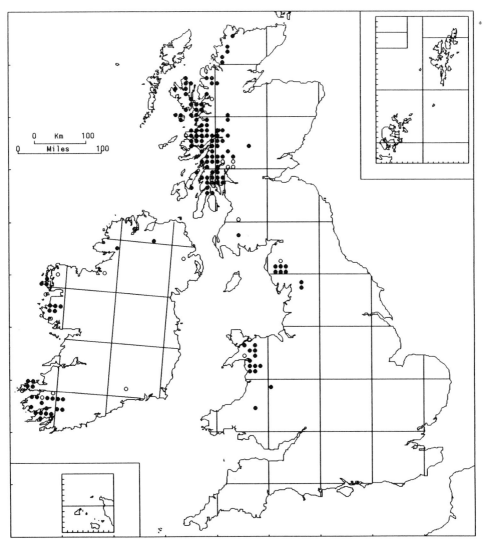

46/9. **Plagiochila exigua** (Tayl.) Tayl. (*P. corniculata* auct.)

A strongly hygrophilous hepatic of deciduous woodland, where it grows usually as pure tufts or patches on moist rock outcrops, boulders, streamside rocks, ravine walls and on trunks and branches of trees, especially ash and hazel. It also grows outside woods, on sheltered rock-faces on steep rocky slopes. 0–610 m (Ballaghbeama Gap). GB 111+10*, IR 30+11*.

Dioecious; male inflorescences frequent, female plants unknown in the British Isles.

Very rare outside the British Isles. S.W. Norway, N.W. France, Pyrenees, Switzerland, Italy. Azores, Canaries, Madeira, southern Appalachians, U.S.A. (a female population), Dominica, Venezuelan Andes. For a map of its European distribution see Bisang *et al.* (1986).

B. AVERIS

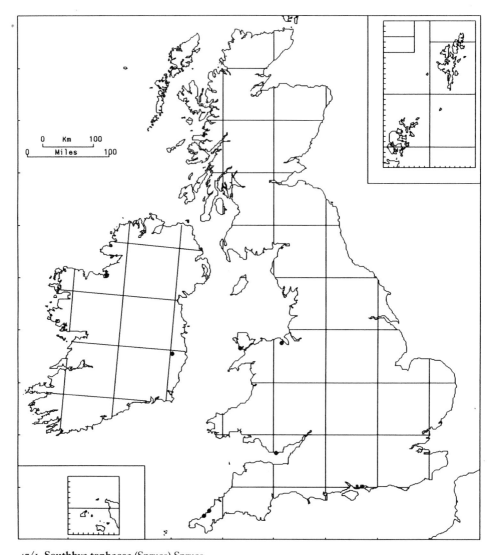

47/1. **Southbya tophacea** (Spruce) Spruce

Found in moist calcareous habitats, including dune-slacks, flushes in or bordering sand-dunes, disused marl-pits, clayey or sandy waste from abandoned mines, wall mortar and dripping tufa on limestone walls and sea-cliffs. It grows in thinly vegetated areas, always associated with *Leiocolea turbinata* and often with *L. badensis*. Lowland. GB 5, IR 2.

Dioecious; sporophytes occasional, maturing from September to April. Reports of gemmae in material from Anglesey cannot be substantiated and must be considered erroneous.

A Mediterranean-Atlantic species, frequent in the Mediterranean countries and reaching its northern limit in the British Isles.

First recognized in Britain in 1961, earlier specimens having been confused with *S. nigrella* (cf. Paton, 1971).

C. D. Preston

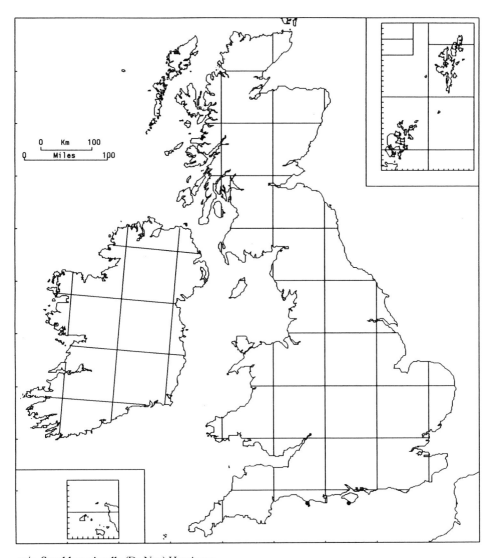

47/2. **Southbya nigrella** (De Not.) Henriques

On thinly vegetated limestone soil in rock crevices, on banks and by tracks near the sea, on stones in disused limestone quarries, and on the friable surface of shaded calcareous rocks, in sites which are moist in winter but dry in summer. The growing season of *S. nigrella* is autumn, winter and early spring. In dry weather the leaves roll inwards towards the stem and the plant becomes almost invisible. Lowland. GB 2.

Paroecious. Sporophytes occasional, maturing in autumn and spring. The species lacks specialized means of vegetative propagation.

A southern species, frequent in the Mediterranean region, extending east to Iraq and U.S.S.R. and at its northern limit in the British Isles.

C. D. Preston

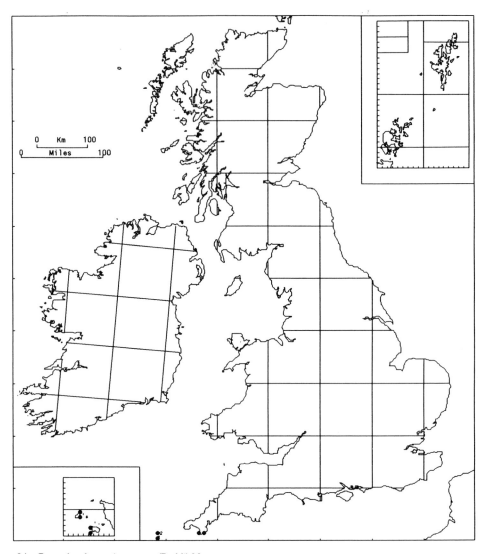

48/1. Gongylanthus ericetorum (Raddi) Nees

In short, open vegetation on shallow soil over granite or serpentine rocks. Found on rocky slopes and in damp hollows in coastal *Calluna* heath, and on roadside banks near the sea, in sites which are damp in winter but subject to summer drought. *Gongylanthus* is calcifuge not only in Britain but throughout its range. It varies in frequency from year to year, and was particularly abundant in the spring of 1977, following the hot summer of 1976 (which reduced competition from vascular plants) and the wet winter of 1976–7. It disappears in summer, presumably surviving the dry season as tuberous stems. Lowland. GB 4.

Dioecious; neither male plants nor sporophytes have been recorded in the British Isles. In the Cameroons sporophytes emerge from a subterranean marsupium and dehisce at the beginning of the wet season, before the new leafy shoots develop (Jones, 1964). The species lacks specialized means of vegetative propagation.

A Mediterranean-Atlantic species, reaching its northern limit in the British Isles. It is also known from the Azores, Madeira, Canaries, W. and S. Africa.

Malloch (1972) gives a detailed description of the habitat of *Gongylanthus* on the Lizard peninsula.

C. D. Preston

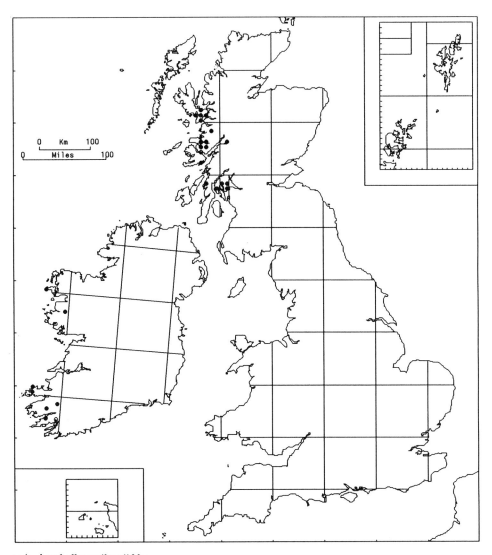

49/1. Acrobolbus wilsonii Nees

It occurs mixed with other bryophytes in thin inconspicuous patches on shady, damp, acid or mildly basic rock-walls of low-lying wooded ravines, or growing epiphytically on *Hymenophyllum wilsonii* or other bryophytes in damp shaded recesses in gorges. It also occurs, more rarely, as dense pure patches on shady sea-cliffs (Skye), and amongst shaded blocks in damp boulder-strewn slopes and block litters (Co. Kerry). It favours N.- or E.-facing localities. Common associates include *Hymenophyllum wilsonii*, several members of the Lejeuneaceae, *Plagiochila exigua* and *Radula* spp. Generally restricted to low altitudes (0–60 m) in Scotland, it ascends to 300 m in S.W. Ireland. GB 17+1*, IR 7+2*.

Dioecious; both sexes occur in the British Isles and sporophytes have been recorded.

It is unknown on the European mainland but has a Macaronesian distribution, occurring in the Azores, Madeira and Canary Islands. Closely allied species occur in S. America.

It is one of the very few epiphyllous bryophytes in the British Isles.

H. J. B. BIRKS

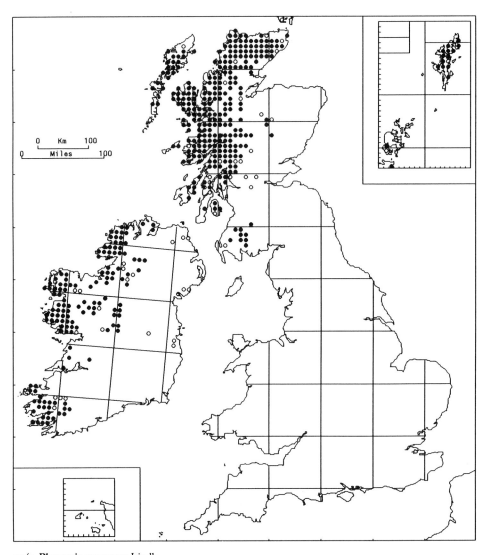

50/1. **Pleurozia purpurea** Lindb.

Locally abundant in the lowlands (0–450 m) on wet, relatively undisturbed ombrotrophic blanket bogs, especially in areas where sphagnum growth is poor or unhealthy. It also occurs on drier, more disturbed blanket bogs and wet heaths. In the uplands, up to about 1000 m, it occurs in oligotrophic heaths, grasslands, and moss, scree and cliff-ledge communities of steep, shaded N.- or E.- facing slopes. It can be an abundant and constant component of the distinctive mixed leafy-liverwort assemblage of dwarf-shrub heaths on steep, block-strewn slopes. In the Cairngorms, this type of community is confined to areas of prolonged snow-cover. GB 283+30*, IR 134+24*.

Monoecious but only female inflorescences known in the British Isles; very rare.

It has a highly disjunct world distribution, with records from S.W. Norway, Faeroes, Jan Mayen, Alaska, the Himalaya and Guadeloupe. Its absence from England and Wales is unaccountable in terms of present-day habitats.

H. J. B. Birks & D. A. Ratcliffe

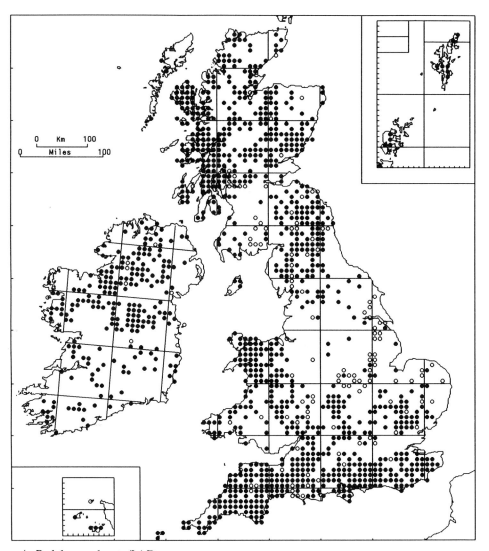

51/1. **Radula complanata** (L.) Dum.

Occurs most commonly as an epiphyte on trunks and roots of elder, ash, beech, elm, hazel and, more rarely, oak, willow and rowan in woods, thickets and hedges. As an epiphyte it avoids deeply shaded sites, and in the drier parts of the British Isles it is most common in wet woods in sheltered localities. Common associates include *Metzgeria furcata*, *Frullania dilatata*, *Lejeunea ulicina* and *Isothecium myosuroides*. It also occurs more locally on shaded basic rocks by streams and in ravines, on rocks in walls in S.W. England, on boulders by lake margins, on shaded sea cliffs, on *Calluna* stems on steep block-strewn slopes, and on mildly basic cliff-faces in the submontane zone. 0–760 m. GB 972 + 118*, IR 221 + 9*.

Paroecious; capsules frequent when growing on trees, rare on rocks. Gemmae common, especially on plants with fruit.

Circumboreal. Widespread in Europe but absent from the far north. South to Macaronesia and N. Africa, and north to S. Greenland.

Probably sensitive to air pollution and as a result has almost certainly declined in parts of C. and N. England.

H. J. B. Birks

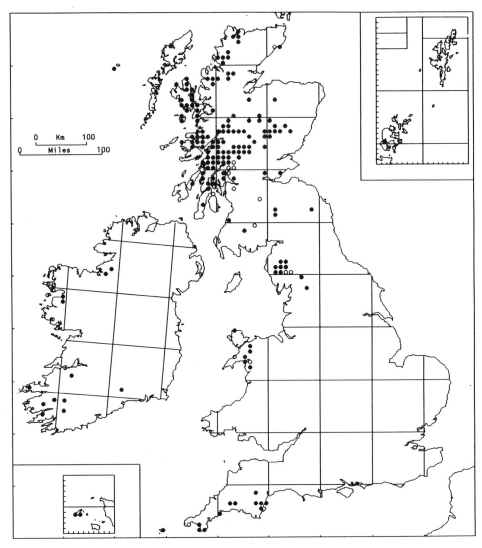

51/2. **Radula lindenbergiana** Gott. ex Hartm. f.

Occurs, usually as small pure but loose patches, on damp shaded mildly basic rocks in both woodland and treeless ravines with *Lejeunea* spp., *Heterocladium heteropterum* and *Hyocomium armoricum*, and as pure, more compact patches on damp shaded basic montane cliffs with *Anoectangium aestivum*, *Distichium capillaceum*, *Fissidens cristatus* and *Pohlia cruda*. It occurs, more rarely, on basic boulders in stable block-screes, on sheltered coastal rocks, on rock outcrops in 'hedges' in Cornwall, and on damp rocks in woodlands. Although commonest above about 300 m, it extends from sea-level to 1200 m. GB 154+14*, IR 12+3*.

Dioecious; female more common than male, sporophytes occasional. Nearly always with gemmae.

Discontinuously circumboreal, south to N. Africa and north to S. Greenland. Widespread but local in Europe, mainly in mountainous areas. Disjunct in southern Africa.

Possibly under-recorded because it cannot be distinguished from *R. complanata* unless gametangia are present. It is probably commoner in Scotland than the map indicates.

H. J. B. BIRKS

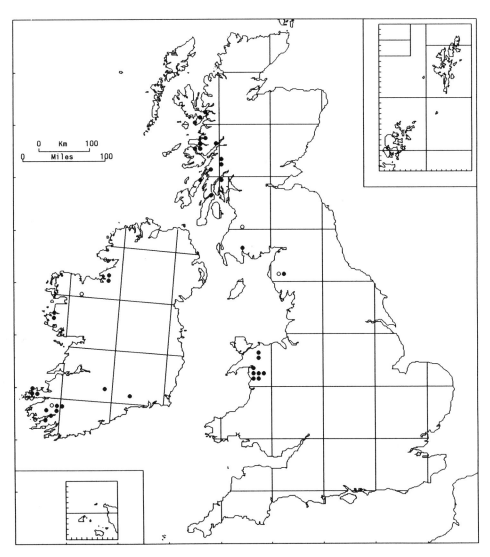

51/3. **Radula voluta** Tayl. ex Gott. & Lindenb.

Despite its rarity it is found in several habitats. It occurs most commonly, but usually in small quantity, growing amongst other bryophytes such as *Hyocomium armoricum* and *Thamnobryum alopecurum* on shaded, damp blocks at and slightly above water-level and near waterfalls and cascades in low-lying wooded ravines and gorges, associated with *Lejeunea lamacerina*, *Riccardia chamedryfolia* and, occasionally, *Jubula hutchinsiae*. More rarely it forms extensive pure yellow-green patches on lightly shaded, mildly basic blocks in and by small streams and cascades in low-lying mixed deciduous woods (mainly oak, elm, ash) and on shaded rock outcrops in woods on steep slopes where there is some intermittent water seepage. It also grows on the sides of caves in cascading wooded streams. In such situations it generally avoids the deeply shaded areas occupied by *Jubula hutchinsiae*. In Wales, the Lake District and Scotland it extends from sea-level to 200 m, but in S.W. Ireland it occurs up to 760 m. GB 24+2*, IR 15+6*.

Dioecious; male plants only.

Unknown elsewhere in Europe. Female plants recorded from the southern Appalachians (U.S.A.).

H. J. B. BIRKS

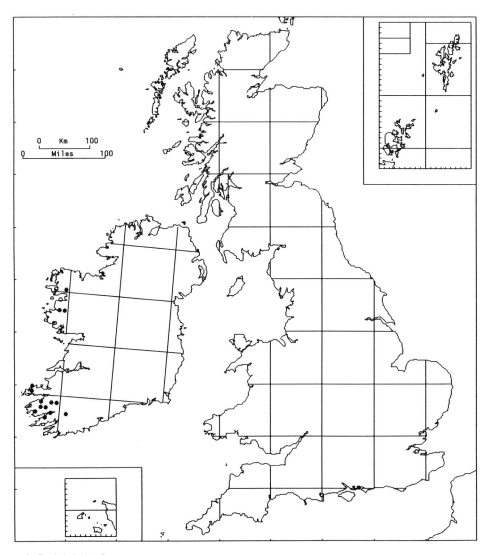

51/4. Radula holtii Spruce

Occurs as pure olive-green patches on deeply shaded, wet rocks in recesses and caves by waterfalls, in dripping ravines, or in the bed of wooded ravines or streams. It occurs more rarely in deep caves between blocks on steep slopes where there is percolating water. Common associates are *Jubula hutchinsiae, Lejeunea holtii, Riccardia chamedryfolia* and, more rarely, *Cyclodictyon laetevirens*. It ascends from sea-level to 330 m in S.W. Ireland. IR 14+1*.

Paroecious; perianths and capsules occasional.

Portugal, Spain. Canaries, Madeira.

Probably commoner in W. Ireland than the map suggests.

H. J. B. Birks

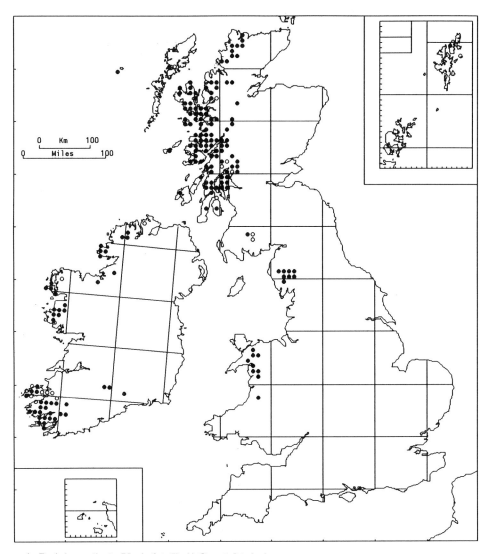

51/5. Radula aquilegia (Hook. f. & Tayl.) Gott. & Lindenb.

Occurs most frequently as pure patches growing on shaded damp acid or mildly basic boulders and vertical rock-walls in low-lying sheltered wooded ravines, often with *Drepanolejeunea hamatifolia*, *Harpalejeunea ovata* and *Lejeunea* spp. creeping through it. It also occurs on sheltered rocks by the sea, on the walls of sea-caves, and in shaded recesses of low sea-cliffs, often with *Frullania microphylla* and *F. teneriffae*. It occur more rarely on damp shaded blocks in deciduous woods with *Adelanthus decipiens*, *Tritomaria exsecta* and *Sematophyllum* spp., and on shaded damp basic cliff-faces, generally N.- or E.-facing, in the submontane or even montane zone with *Preissia quadrata*, *Anoectangium aestivum* and *Pohlia cruda*. In extremely humid places such as the spray zone of waterfalls and on rocky N.- or E.-facing slopes in high rainfall areas of W. Scotland and S.W. Ireland it also grows mixed with other hepatics as an epiphyte on oak, hazel and, more rarely, birch trunks and lower branches. In its English and Welsh localities it is restricted to mildly basic substrata. It ascends from sea-level to at least 920 m. GB 126+11*, IR 47+12*.

Dioecious; males uncommon, females common; perianths and capsules very rare.

Faeroes, S.W. Norway, Spain, Portugal. Azores, Canaries, Madeira, Himalaya.

H. J. B. Birks and D. A. Ratcliffe

239

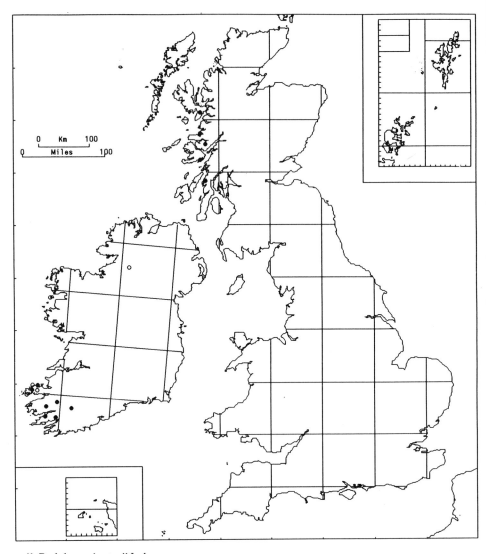

51/6. **Radula carringtonii** Jack

Occurs in small quantity as compact, olive-brown patches on shaded mildly basic blocks and on damp, shaded vertical rock-walls in low-lying wooded ravines, often near the sea. Associates include *Drepanolejeunea hamatifolia, Harpalejeunea ovata, Plagiochila exigua, Radula aquilegia* and, more rarely, *Acrobolbus wilsonii*. All Scottish localities are below 100 m, but in S.W. Ireland it extends to 350 m. GB 4+1*, IR 7+5*.

Dioecious. Both sexes occur in the British Isles; sporophytes are unknown.

Outside the British Isles it is confined to N.W. Spain, Madeira and Azores. The northernmost known world localities are in Scotland.

It is often inconspicuous and is easily overlooked. In view of the availability of seemingly suitable habitats in W. Scotland and S.W. Ireland it is probably commoner than the map suggests.

H. J. B. Birks

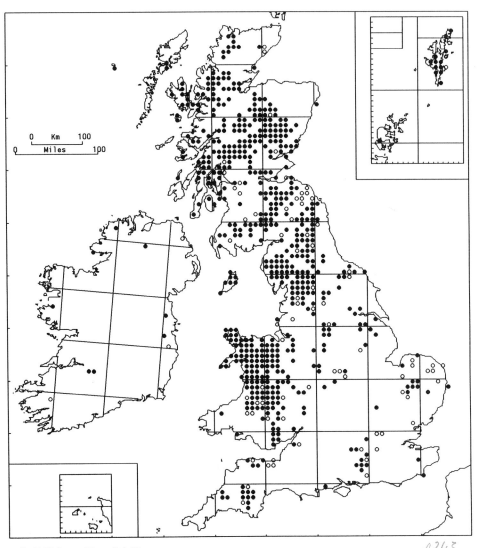

p 243

52/1. Ptilidium ciliare (L.) Hampe

A very wide-ranging species in the uplands of Wales, N. England and Scotland, extending through submontane heath and grassland, blanket mire, screes and crags, to high montane moss and lichen heath. It is generally less frequent in the lowlands where habitats include dwarf-shrub, grass and dune heath, as well as heathy birch and pine woodland. It usually occurs in scattered patches amongst other bryophytes, on soil, around the base of outcrops or on rock ledges; mostly, but not exclusively, on well-drained base-poor substrata. 0–1300 m (summit plateau of Ben Nevis). GB 581+78*, IR 8+5*.

Dioecious; sporophytes are very rare.

Widely distributed in the boreal and montane regions of Europe. Elsewhere it occurs in Asia, N. America, southern S. America and New Zealand.

The scarcity of Irish records of this conspicuous, attractive species is notable; it was found as far back as the first half of the 19th century in Co. Kerry. However, it is a species with somewhat continental tendencies, strikingly more abundant in the eastern Highlands of Scotland than in the western.

T. H. BLACKSTOCK

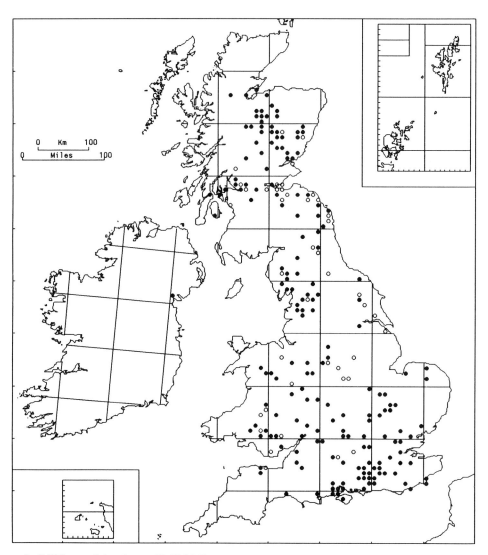

52/2. Ptilidium pulcherrimum (G. Web.) Vainio

This predominantly corticolous species occurs mainly in semi-natural woodlands, occasionally in plantations and scrub. Although often present in small quantity, it is able to colonize a variety of tree species, including birch, ash and less frequently oak in drier woodland, and willow and alder in carr and damp woods; other hosts include shrubs such as elder, and conifers including pine and juniper. It is also found on rotting wood and stumps, and more rarely on rocks. Mainly lowland, to 400 m in N.E. Scotland (River Findhorn). GB 182+44*, IR 1.

Dioecious; sporophytes are seldom reported in Britain but may be fairly frequent. They are produced regularly in a birchwood in Glen Doll, Angus (Duncan, 1966). It is known to be highly fertile in N. America and some parts of Europe. There is no specialized method of vegetative dispersal.

Circumboreal. Widely distributed in the boreal and montane regions of Europe, extending eastwards to Russia.

A high proportion of the records are relatively recent (post–1950), and it has been suggested by Wallace (1963) that *P. pulcherrimum* may be spreading in Britain. It was first recorded in Ireland in 1953.

T. H. BLACKSTOCK

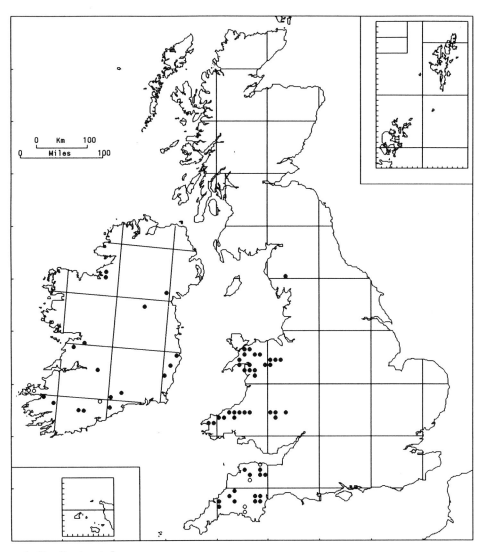

53/1. Porella pinnata L.

Rare but locally abundant on rocks and, less frequently, tree-roots at or slightly above water-level in low-lying wooded fast-flowing rivers and large streams, growing with *Lejeunea lamacerina*, *Hyocomium armoricum* and *Thamnobryum alopecurum*. It occurs more rarely submerged on boulders by lake margins. It is confined to non-calcareous waters but is also absent from extremely acid waters. 0–200 m. GB 46+4*, IR 17+4*.

Dioecious; only female plants occur in the British Isles.

S. and W. Europe. Madeira, N. Africa and N. and C. America. The British and Irish localities are its northernmost in Europe.

H. J. B. BIRKS

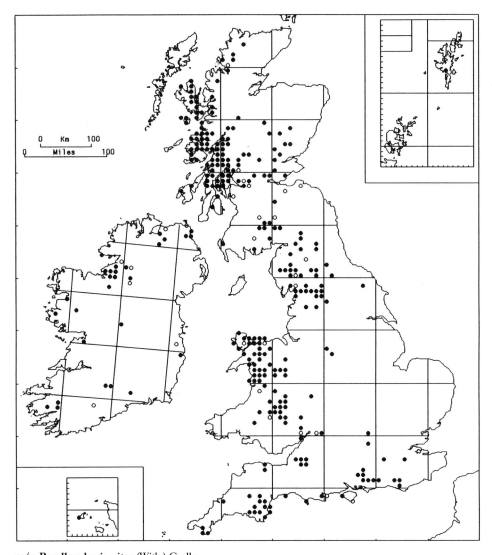

53/2. **Porella arboris-vitae** (With.) Grolle

Widespread but rarely present in large amounts. On shaded dry boulders and vertical rock-faces in sheltered low-lying wooded ravines, generally on base-rich substrates, and often growing with other bryophytes such as *Frullania tamarisci* and *Plagiochila porelloides*. It also occurs on shaded basic montane cliffs and, more rarely, as an epiphyte on ash and hazel in wooded ravines, on shaded basic rocks in woods, on tree-bases, and in grassy turf, either on N.-facing chalk slopes or steep, shaded slopes below basic cliffs. More shade-demanding than *P. platyphylla*. 0–610 m. GB 255+22*, IR 28+8*.

Dioecious; only female known; sex organs common.

Widespread in Europe, but absent from the north.

No attempt has been made to map var. *killarniensis* (Pears.) Corley or var. *obscura* (Nees) Corley.

H. J. B. BIRKS

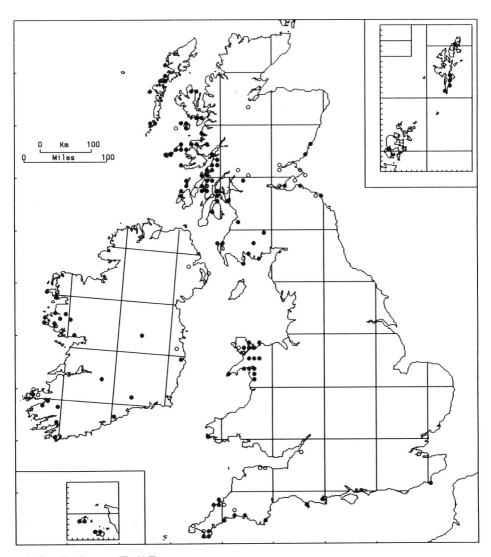

53/3. **Porella obtusata** (Tayl.) Trev.

On dry, rather exposed S.- or W.-facing acid or basic low-lying rocks, often near the sea, growing with *Ptero-gonium gracile*, *Ptychomitrium polyphyllum* and *Ulota phyllantha*, sometimes just above high-water mark. It also occurs, more rarely, as an epiphyte on ash, and on sheltered rocks in lanes, heaths, woodland, and 'hedges' in S.W. England. 0–400 m. GB 94+24*, IR 21+6*.

Dioecious; females common, male plants and capsules unknown in the British Isles.

Mediterranean-Atlantic distribution in Europe, extending from Yugoslavia, Italy and Corsica along the western seaboard to S.W. Norway. Also in Azores, Canaries, Madeira, N. Africa, C. Asia.

H. J. B. BIRKS

245

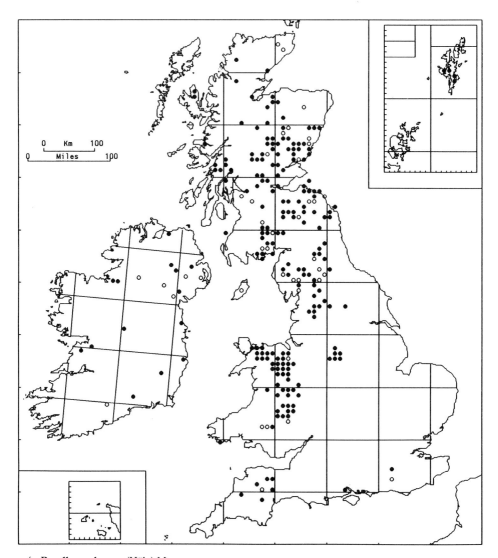

53/4. Porella cordaeana (Hüb.) Moore

Curiously patchy in its occurrence and abundance, it is most common on silty tree-bases and roots, rocks, and stones by low-lying streams and rivers, often at or near water-level or in the flood zone. It also occurs, more rarely, on damp shaded basic rocks in low-lying wooded ravines, on shaded basic rocks in the montane zone, and submerged, or just above the water-level, on large boulders beside lakes. It favours basic substrata but is rare on limestone in contrast to *P. platyphylla*. Mainly a lowland species (0–400 m), it extends to 1000 m in the Breadalbane mountains. GB 193+34*, IR 14+6*.

Dioecious; fruit rare.

Widespread in Europe; absent from the extreme north.

Var. *cordaeana* and var. *simplicior* (Zett.) S. Arn. have not been separated for mapping purposes. Var. *faeroensis* (C. Jens.) E. W. Jones has its only British locality at Sand Water, Bixter, Shetland. The species is possibly commoner than the map indicates.

H. J. B. Birks

246

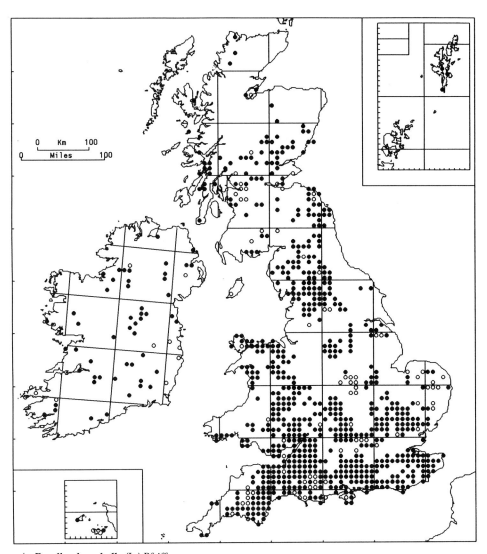

53/5. Porella platyphylla (L.) Pfeiff.

In S. England it is commonest on and around tree-bases and roots, particularly beech and ash, in woodland on chalk soils, and on bare chalk soil on wooded slopes and shaded banks, with *Anomodon viticulosus, Ctenidium molluscum* and *Encalypta streptocarpa*. It also occurs, more rarely, on logs. Further north and west it is commonest on limestone or other basic rocks and stone walls, usually where there is some light shade. It also occurs on tree-bases by streams and in ravines, and in shady lanes and hedges. In its few N. Scottish localities it is mainly on dry but shaded limestone cliffs with *Anomodon viticulosus, Encalypta streptocarpa* and *Tortula intermedia*. It is less shade-demanding than *P. arboris-vitae* but more strongly calcicolous and less moisture-demanding than either *P. arboris-vitae* or *P. cordaeana*. Mainly lowland but extends to 600 m in the Craven Pennines. GB 686+81*, IR 51+11*.

Dioecious; fruit rare.

Widely distributed in the Northern Hemisphere, south to N. Africa, C. Asia and California; absent from the Arctic and most of the boreal zone.

H. J. B. Birks

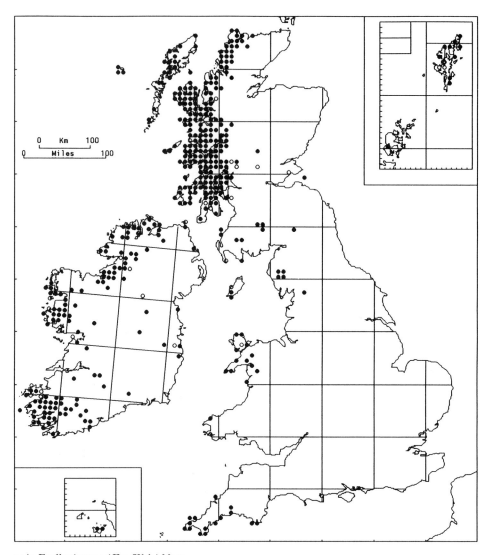

54/1. **Frullania teneriffae** (Web.) Nees

Most frequent on dry siliceous rocks near the sea, often in exposed situations with *Schistidium maritimum*, *Ulota phyllantha*, *Anaptychia fusca* and *Ramalina siliquosa*, and as an epiphyte on trunks and lower branches of hazel, birch, oak, ash, and, more rarely, elm, rowan and willow in woods, generally on steep blocky slopes. It also occurs on shaded boulders and mildly basic rock-faces in low-lying wooded ravines, on blocks in mixed deciduous woods on steep slopes, in mixed *Calluna-Vaccinium* dwarf shrub-heaths on N.- or E.-facing block-strewn slopes, and on treeless but shady cliff-faces and in gullies above about 300 m. In the extreme west it also occurs on dry, rather exposed rocks in inland block-litters. It is strikingly abundant on the sea-spray-influenced islands of the St Kilda group, growing in all rock habitats, grasslands and heaths, and on banks and heather stems. Occurs on a wide range of rock types in W. Scotland and Ireland, but favours mildly basic rocks in its eastern and southern localities. 0–700 m. GB 272+16*, IR 120+13*.

Dioecious; female plants common and capsules frequent.

Faeroes, Portugal, France, Spain, Sardinia. Morocco, Azores, Canaries, Madeira.

H. J. B. Birks and D. A. Ratcliffe

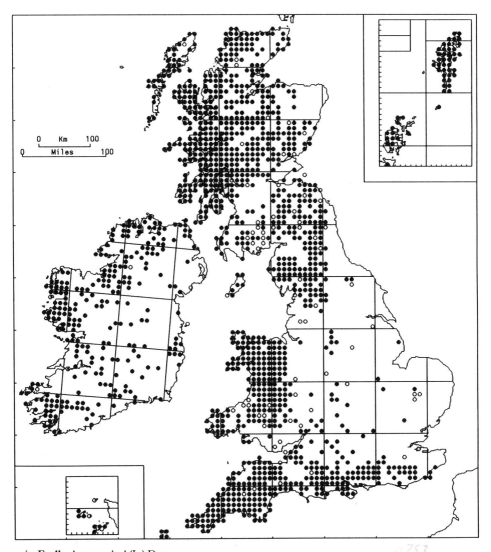

54/2. Frullania tamarisci (L.) Dum.

Common and locally abundant in several habitats – mossy boulders in deciduous woods; shaded rocks by streams, in ravines, and by the sea; damp turf and grassland both in the lowlands and uplands, often on basic or calcareous soils; dry, often exposed boulders in stable block-litters; stone walls; trunks and lower branches of trees (particularly oak, hazel, ash, birch) and stems of heather and gorse; rotten logs; dwarf-shrub heaths; cliff-faces and gullies; shaded peat and earthy banks; and short dune-turf. Generally requires some shade or shelter, except in the extreme west of Scotland, and commonly favours mildly basic substrates outside the extreme west. Commonest below 600 m but extending to 1000 m in the Scottish Highlands. GB 1142+74*, IR 256+5*.

Dioecious; female plants frequent, capsules occasional.

Widespread in Europe.

H. J. B. BIRKS

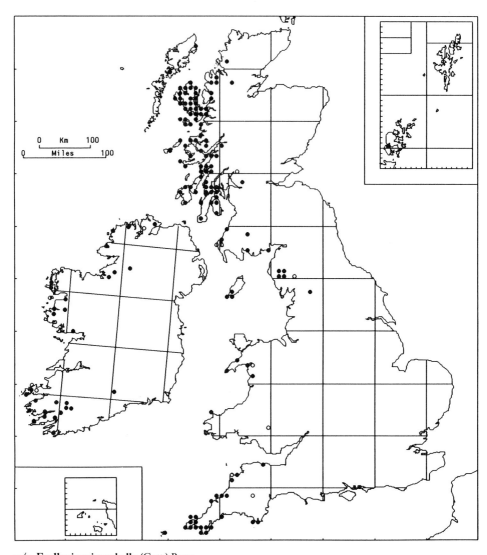

54/3. **Frullania microphylla** (Gott.) Pears.

On near-vertical, seasonally moist, shaded mildly basic rocks by the sea, often growing with *Frullania teneriffae*, in sheltered sea-caves with *Lophocolea fragrans* and *Radula aquilegia*, and, more rarely, on shaded rocks in woods, shaded vertical rock-walls in low-lying wooded ravines, on blocks by lakes, and in treeless but shady cliff-faces and in gullies above about 300 m. 0–370 m in Scotland, extending to 610 m in Co. Kerry. GB 124+9*, IR 29+4*.

Dioecious; female plants frequent, capsules occasional.

Faeroes, France, Germany, Spain, Portugal. Azores, N. Africa.

H. J. B. Birks

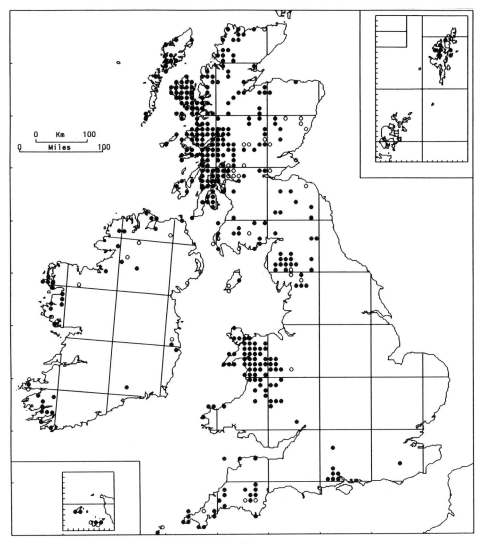

54/4. Frullania fragilifolia (Tayl.) Gott., Lindenb. & Nees

On dry, exposed, acid or mildly basic, often igneous, rocks near the sea with *Frullania teneriffae, Ulota hutchinsiae* and *U. phyllantha*, and on dry basic igneous rock-outcrops and cliffs at low altitudes with *Grimmia trichophylla, Ptychomitrium polyphyllum, Schistidium apocarpum* and *Tortula subulata*. It also occurs on dry or slightly irrigated, shaded vertical rock-walls in low-lying wooded ravines and on rocks by wooded streams, on basic cliff-faces and in shaded gullies in the uplands, on shaded rocks in woods, and as an epiphyte on trunks of oak, ash, hazel, and sycamore with *Orthotrichum* spp. and *Zygodon* spp. in open woodland and hedgerows. Mainly a lowland species (0–400 m), it ascends to 850 m in the C. Highlands. GB 347+34*, IR 35+9*.

Dioecious; female plants frequent, capsules rare.

Widespread but local in Europe, extending from the Mediterranean to Finland and Iceland. Macaronesia. Possibly commoner in parts of W. Scotland and in Ireland than the map indicates.

H. J. B. BIRKS

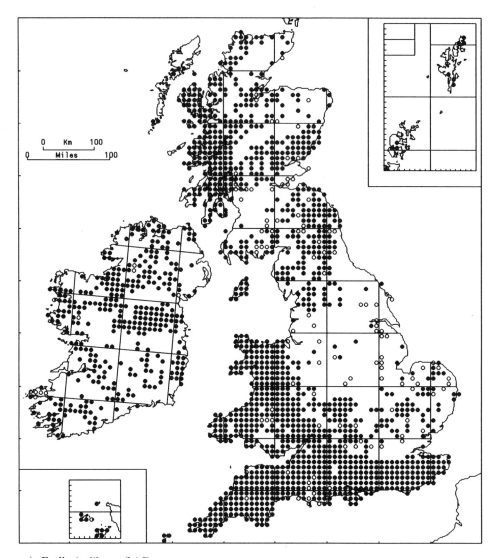

54/5. Frullania dilatata (L.) Dum.

Most frequent as an epiphyte on lower branches and trunks of hazel, ash, elm, willow, elder, oak, and, more rarely, rowan, alder, blackthorn, sycamore, beech, and birch in open woods, plantations and hedgerows, growing with *Metzgeria furcata*, *Radula complanata*, *Hypnum cupressiforme* var. *resupinatum* and Orthotrichaceae. It also occurs on gorse and heather stems in the west. Usually in sheltered and rather dry situations. Also on shaded rocks in woods and wooded ravines, stonewalls, roadside rocks, tombstones, buildings, rocks near the sea, and on fallen logs. Usually below 310 m, but extending to 600 m in the Scottish Highlands. GB 1298+110*, IR 334+10*.

Dioecious; perianths and capsules common.

Widespread throughout Europe, except in the extreme north. Extending eastwards to Siberia and China. Absent from N. America.

Sensitive to atmospheric pollution and consequently rare in the Midlands and parts of N.W. England.

H. J. B. BIRKS

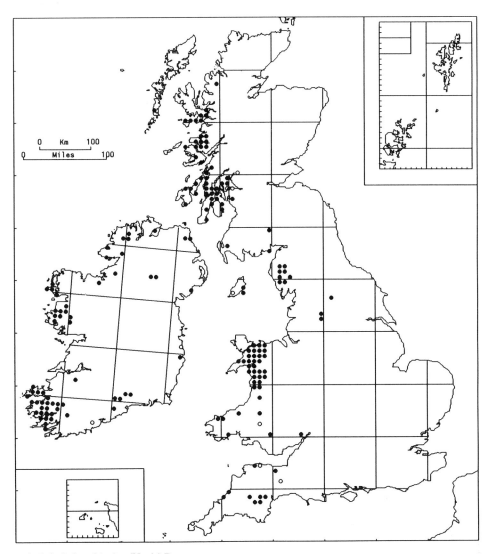

55/1. Jubula hutchinsiae (Hook.) Dum.

Most commonly as pure, dark-green patches on shaded, wet, often constantly dripping rocks in low-lying wooded ravines, especially around waterfalls and cascades and in deeply shaded humid recesses in caves formed from jammed blocks in the bed of ravines. It also occurs on wet shaded stream banks in wooded glens, particularly in S. Ireland and S.W. England, and in damp, shaded crevices and caves in raised-beach cliffs along the west coast of Scotland. It sometimes occurs epiphytically on the fronds of *Trichomanes speciosum* and on the stems of *Thamnobryum alopecurum* in dripping caves by waterfalls and in wooded ravines. It occurs on a wide range of rock types including limestone (e.g. in caves in Co. Leitrim) but is very rare on the most acidic rocks. Outside Ireland, it is confined to low altitudes (0–250 m), but in Co. Kerry it reaches 610 m. GB 109+8*, IR 60+2*.

Monoecious; inflorescences and capsules occasional.

Ssp. *hutchinsiae* is recorded from France, Spain and Macaronesia. Ssp. *javanica* (Steph.) Verd. occurs in Turkey, Caucasus, Melanesia, and Asia; other subspecies are recorded from N. and S. America.

H. J. B. Birks

253

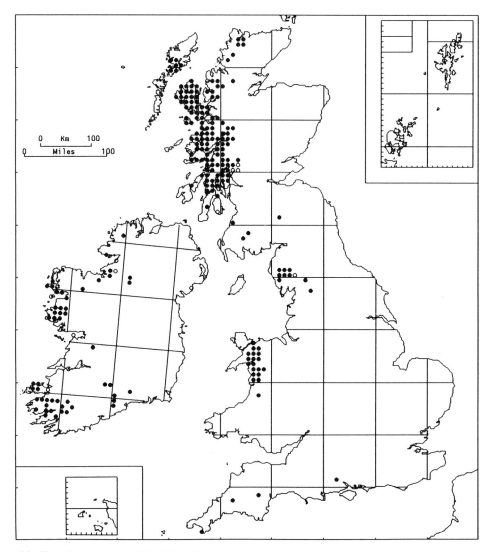

56/1. **Harpalejeunea ovata** (Hook.) Schiffn.

Occurs on damp, lightly-shaded, often vertical rock-walls of low-lying wooded ravines, either as pure patches growing directly on the rock or as scattered stems growing amidst *F. tamarisci, F. teneriffae, Lejeunea cavifolia, L. patens* or *Metzgeria* spp. Also occurs on damp but not wet boulders in ravines, by waterfalls, and in and near wooded streams, on rocks in stable block-litters on steep N.- or E.-facing slopes, on shaded walls of sea-caves and sheltered rocks by the sea, on damp N.- or E.-facing montane cliffs and crags, and in shaded gullies in montane cliffs. Rare as an epiphyte on birch, oak, ash, hazel, heather, and, in the New Forest, beech. Tends to favour mildly basic substrates. Common associates include *Aphanolejeunea microscopica, Drepanolejeunea hamatifolia* and *Lejeunea* spp. Can also occur in drier and less shaded situations than *A. microscopica* and *D. hamatifolia*, but is then usually on more basic substrates with *Cololejeunea calcarea* and *Lejeunea cavifolia*. 0–550 m, extending to 600 m in S.W. Ireland. GB 165+6*, IR 50+7*.

Dioecious; female plants frequent.

Faeroes, S.W. Norway, N.W. France, Spain, Portugal, Corsica, Italy, Switzerland. Canaries, Madeira, Azores. For a map of its European distribution, see Bisang *et al.* (1986).

H. J. B. BIRKS

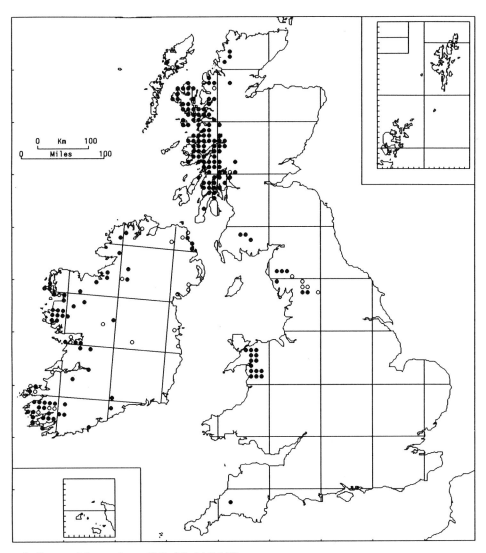

57/1. Drepanolejeunea hamatifolia (Hook.) Schiffn.

On damp, shaded, often vertical, rock-walls of sheltered, low-lying wooded ravines and by waterfalls, usually as pure thin patches growing directly on the rock, or, more rarely, growing over other bryophytes, particularly *Frullania* spp. and *Radula aquilegia*. It also occurs on damp but not wet, shaded boulders in ravines and on rocks in wooded streams, on shaded walls of sea-caves and sheltered rocks by the sea, on boulders in stable block-litters on steep N.- or E.-facing slopes, and in damp N.- or E.-facing gullies or basic cliffs in the uplands. In the extreme west it also occurs epiphytically on birch, oak, hazel, sallow, ash and heather, often on trees and bushes growing in or at the edge of ravines and by waterfalls. Common associates include *Aphanolejeunea microscopica, Colura calyptrifolia, Harpalejeunea ovata* and *Lejeunea* spp. In the lowlands it is seemingly indifferent to rock type, but above about 200 m and at the edge of its range it is virtually confined to basic substrates. Generally more base-demanding than *A. microscopica*, more shade-demanding than *H. ovata*, and commoner than *C. calyptrifolia*. Mainly in the lowlands (0–430 m), extending to 610 m in S.W. Ireland. GB 128+10*, IR 61+24*.

Monoecious; male inflorescences, perianths and fruit rather rare.

N.W. France, Pyrenees. Azores, Madeira, tropical and southern Africa.

H. J. B. BIRKS

255

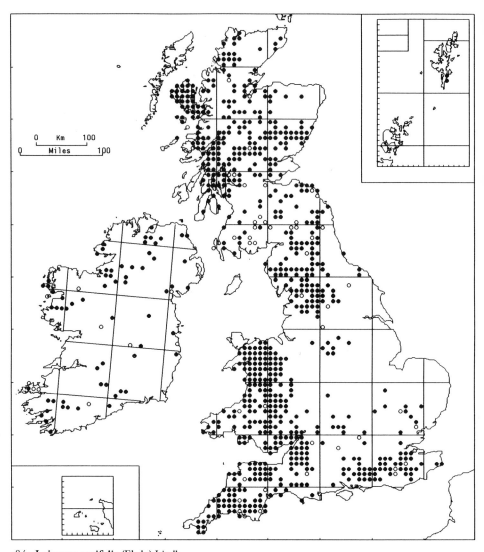

58/1. **Lejeunea cavifolia** (Ehrh.) Lindb.

On damp, shaded basic rock-faces, boulders, and low cliffs in woods and wooded and treeless ravines, often growing among other bryophytes such as *Frullania tamarisci*, *Metzgeria furcata* and *Plagiochila porelloides*, on shaded basic montane cliff-faces and gullies, on sheltered sea cliffs, on and amongst blocks and in shaded humid recesses in stable block-litters, and on shaded soil banks in wooded valleys and by wooded streams. More rarely epiphytic on ash or hazel. It is more base-demanding than *L. lamacerina* or *L. patens* and less shade-demanding than *L. lamacerina*. Often associated with *Cololejeunea calcarea*, *Marchesinia mackaii*, *Ctenidium molluscum* and *Fissidens cristatus*. 0–1020 m (Aonach Beag). GB 683+49*, IR 64+10*.

Monoecious; perianths common; fruit rare, commonest when growing epiphytically.

Widespread but local in Europe, except the extreme north and east. Azores, Canaries, Madeira, N. Africa, Asia, N. America.

H. J. B. BIRKS

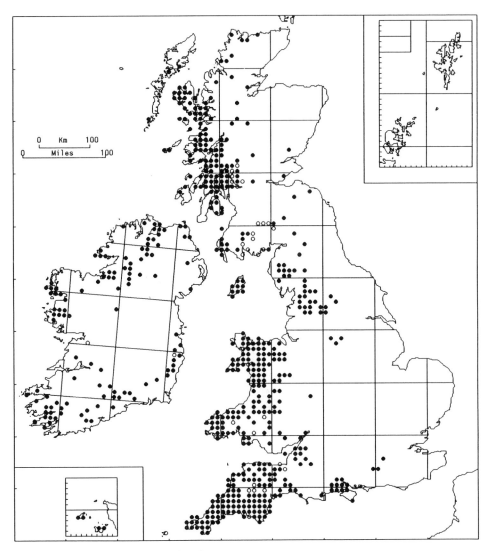

58/2. **Lejeunea lamacerina** (Steph.) Schiffn.

Most common growing over other bryophytes (e.g. *Heterocladium heteropterum*, *Hyocomium armoricum* and *Thamnobryum alopecurum*) in damp, shaded situations such as at or near water-level on wet boulders and tree-roots in wooded valleys and streams, on moist, sometimes dripping rock-walls in damp, humid, wooded ravines and by waterfalls, and in shaded recesses between boulders in cascades and on damp, block-strewn slopes where water percolates. Also in dripping sea-caves, on sheltered rocks by the sea, on shaded bridge-supports in wooded river valleys, on trees and decaying logs in woods, and on soil banks by shaded streams. Associates include *Lophocolea fragrans*, *Riccardia chamedryfolia* and, more rarely, *Jubula hutchinsiae*, *Porella pinnata*, *Fissidens curnovii*, *F. polyphyllus* and *F. rivularis*. Generally in damper, more shaded situations than *L. cavifolia* or *L. patens* and less base-demanding than *L. cavifolia*. Mainly in the lowlands (0–150 m), extending to 650 m in S.W. Ireland. GB 447+31*, IR 109+2*.

Monoecious; fruit rare.

Faeroes, France, Spain, Italy. Azores, Madeira, Canaries.

Probably commoner in the extreme west than the map indicates.

H. J .B. BIRKS

257

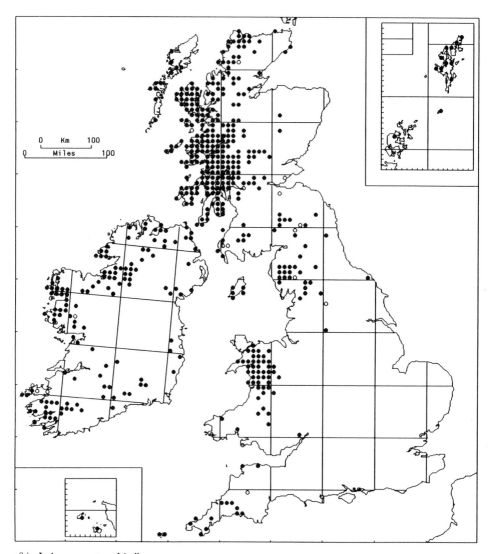

58/3. **Lejeunea patens** Lindb.

Grows as pure patches or amongst other bryophytes on moist, sheltered, mildly basic rock-faces and boulders in wooded or treeless ravines and gullies, on damp, periodically flushed rocks in woods on steep, block-strewn slopes, on sheltered sea-cliffs, on shaded montane cliff-faces, and on and amongst boulders and in shaded recesses in stable block-screes on N.- or E.-facing slopes. More rarely as an epiphyte on trees and heather stems in damp, sheltered situations, often by ravines, and on logs in wooded valleys and shaded banks. It is indifferent to substrate but avoids the most sterile acid rocks and is very rare on limestone. Generally less base-demanding than *L. cavifolia* and less shade-demanding than *L. lamacerina*. Frequent associates include *Harpalejeunea ovata*, *Metzgeria conjugata*, *Plagiochila spinulosa*, *Porella arboris-vitae*, *Radula aquilegia* and *R. lindenbergiana*. 0–800 m (Coire Cheap). GB 376+15*, IR 128+7*.

Monoecious; perianths common, fruit rare.

W. Europe from Norway and Faeroes south to Spain, Balearics, and Macaronesia and east to N. Italy, Germany, S.E. Yugoslavia and Crimea. Caucasus, S.W. Asia.

H. J. B. Birks

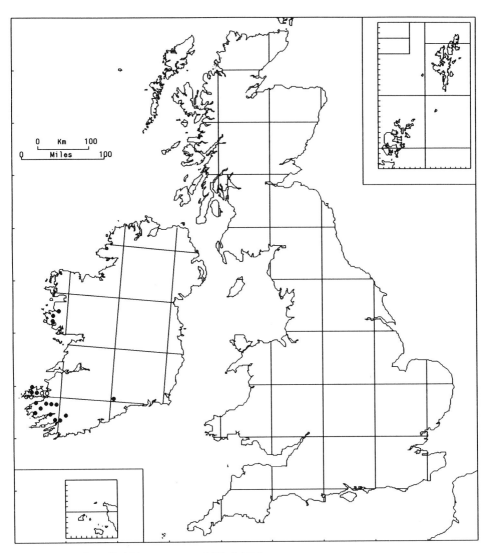

58/4. **Lejeunea flava** (Sw.) Nees ssp. **moorei** (Lindb.) Schust.

On dry, lightly shaded, mildly basic boulders, rock outcrops, and periodically flushed blocks in deciduous woods and wooded glens, growing as pure patches or mixed with *Adelanthus decipiens*, *Plagiochila spinulosa*, *Scapania gracilis* and *Dicranum scottianum*. It also occurs on dry but shaded blocks amongst cascading streams, often with little or no tree-cover, in block-litters on N.- or E.-facing slopes, growing with *Plagiochila exigua* and *Radula aquilegia*, and on a shaded drystone wall by a wooded valley in Co. Waterford. 0–300 m. IR 17+4*.

Monoecious; male inflorescences and female branches common, fruit not known in Ireland.

Azores, Madeira, Canaries. Other subspecies occur in the tropics, Australasia, N. America, S. America.

H. J. B. Birks

259

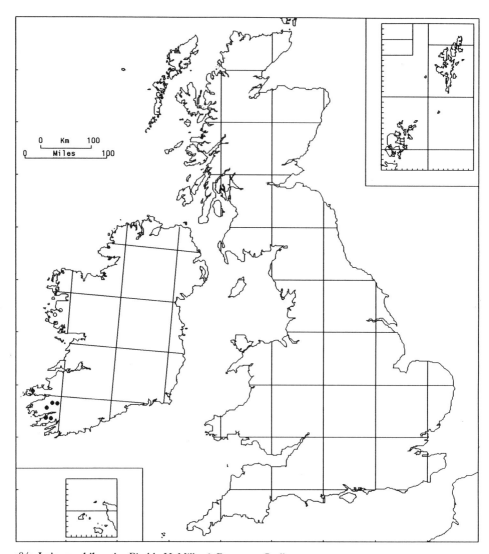

58/5. Lejeunea hibernica Bischl., H. Miller & Bonner ex Grolle

Occurs as scattered stems growing amongst *Metzgeria* spp. and *Radula* spp. (particularly *R. aquilegia* and *R. carringtonii*) on damp shaded boulders in sheltered gullies, cascades, block-litters, wooded glens and ravines; also in wet shaded crevices in cliffs, usually associated with *Drepanolejeunea hamatifolia*, *Harpalejeunea ovata*, *Plagiochila exigua* and, occasionally, *Marchesinia mackaii*. It favours mildly basic rocks. 0–370 m (Mullaghanattin). IR 6+2*.

Autoecious; gametangia fairly frequent, capsules not known in Ireland.

Azores, Canaries, Madeira.

Probably commoner in W. Ireland than the map indicates, as it is small, inconspicuous, and often difficult to spot when it and its associates are wet.

H. J. B. Birks

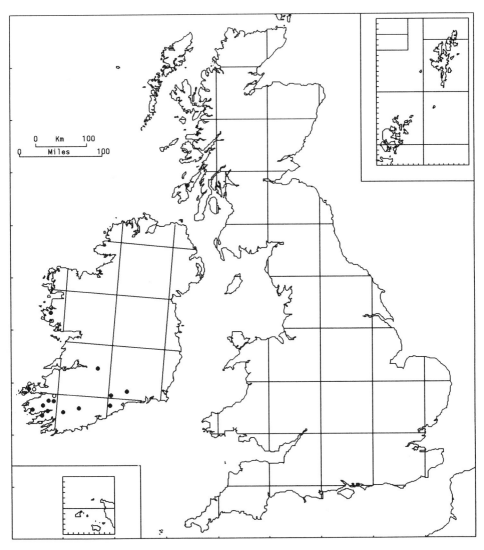

58/6. Lejeunea holtii Spruce

Occurs as pure patches or mixed with other bryophytes on wet, shaded rocks in wooded glens and ravines, in dripping, shaded caves by waterfalls, on blocks in cascades, and in wet, dripping caves along the raised-beach platform of Islay. More rarely on stipes and fronds of senescent *Trichomanes speciosum* in deeply shaded, dripping caves and recesses. Commonly associated with *Jubula hutchinsiae*, *Radula holtii*, *Riccardia chamedryfolia*, *Cyclodictyon laetevirens* and *Thamnobryum alopecurum*. It favours slightly basic substrates and with increasing elevation (above 150 m) becomes restricted to more basic, deeply shaded, and dripping situations. 0–330 m. GB 1, IR 14+5*.

Monoecious; male inflorescences and perianths frequent, sporophytes rare.

Spain. Azores, Madeira, Canaries.

H. J. B. BIRKS

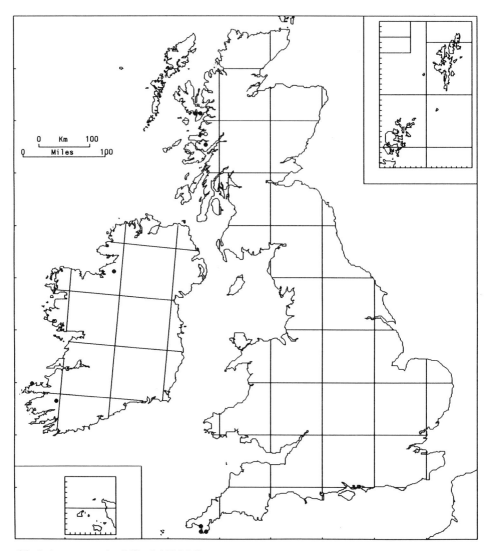

58/7. **Lejeunea mandonii** (Steph.) K. Müll.

On shaded but dry or periodically irrigated N.- or E.- facing rocks or cliffs in wooded ravines, sheltered sea caves or woods; on shaded N.-facing limestone cliffs; and on sheltered coastal rocks. It frequently grows with other hepatics such as *Frullania* spp., *Lejeunea* spp., *Marchesinia mackaii* and *Radula* spp. The substrate is invariably basic (limestone, serpentine, calcareous Old Red Sandstone, schist). It also occurs, more rarely, growing with other hepatics on elm and ash trees and on rotting logs in shaded ravines. Near sea-level in its Scottish and English localities, extending to 360 m in Co. Leitrim. GB 6+1*, IR 3.

Monoecious; perianths common, sporophytes frequent.

Spain, Portugal. Canaries, Madeira.

Probably under-recorded, but undoubtedly a rare plant with, for example, only two known localities on Skye despite an abundance of seemingly suitable habitats and much searching. The Scottish localities are its northernmost world stations.

H. J. B. Birks

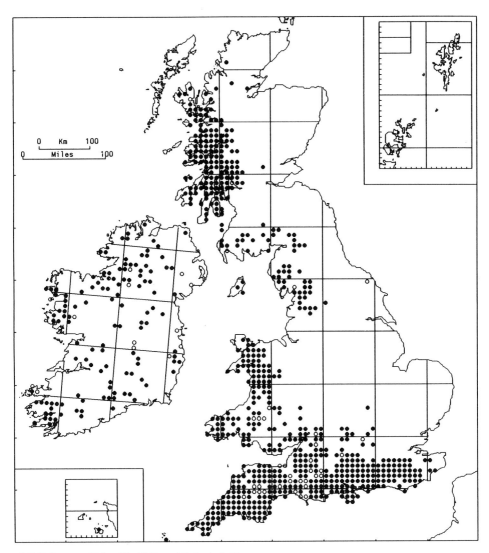

58/8. **Lejeunea ulicina** (Tayl.) Gott., Lindenb. & Nees

Epiphytic on trunks and lower branches of birch, alder, oak, beech and willow, less common on rowan, hazel, ash, elm, elder, and occasional on sycamore, rhododendron, apple, field maple, hawthorn, blackthorn, privet and pine in woods, ravines, gullies and swamp carrs, and by roadsides and in hedges. Often associated with *Frullania dilatata*, *Metzgeria furcata* and *Ulota* spp. Also on stems of heather and gorse. More rarely on shaded acid or mildly basic rocks in ravines, on sheltered rocks by the sea, and peaty banks on damp, rocky slopes. 0–240 m. GB 610+37*, IR 124+13*.

Dioecious; female plants frequent, male plants unknown in Britain and Ireland.

Scattered throughout S. and W. Europe, extending from the Mediterranean (Corsica, Italy, Yugoslavia) through C. Europe (Austria, Switzerland) and W. Europe to Norway. Canaries, Azores, Madeira, Tunisia, tropical and southern Africa, Asia, N. America.

H. J. B. BIRKS

263

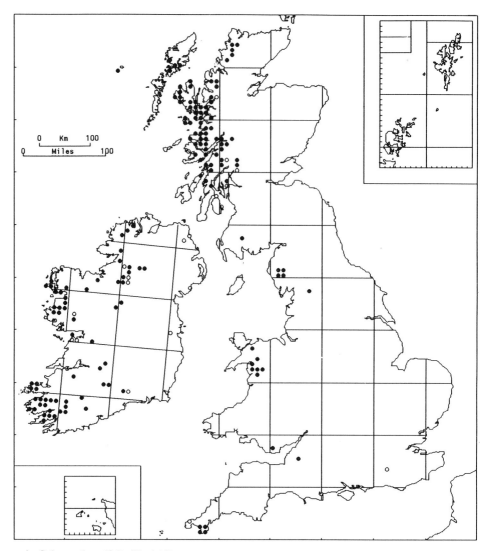

59/1. **Colura calyptrifolia** (Hook.) Dum.

It grows in small, dense, pure patches or as scattered stems amongst *Frullania* spp. or *Radula aquilegia* on damp, shaded rocks and boulders in humid situations, including sheltered ravines, rocky gorges, low-lying wooded streamsides, lakesides, rocky coasts, and damp N.- or E.-facing montane cliffs. It also occurs epiphytically on heather stems on steep, shaded block-strewn slopes; on gorse, *Erica* and willow stems in S.W. England and S. Wales; and on conifer trunks. In rupestral habitats it is commonly associated with *Drepanolejeunea hamatifolia* and *Radula aquilegia*. It tends to favour mildly basic rocks except in S.W. Ireland where it is seemingly indifferent to rock type. Outside S.W. England, it is invariably but inexplicably the rarest of the quartet *Aphanolejeunea*, *Drepanolejeunea*, *Harpalejeunea* and *Colura*. 0–460 m, extending to 610 m in W. Ireland. GB 90+10*, IR 62+11*.

Monoecious; perianths usually present, capsules rare. Gemmae occasional.

N.W. France. Azores, Kenya (Mt. Kilimanjaro), S. America.

H. J. B. BIRKS

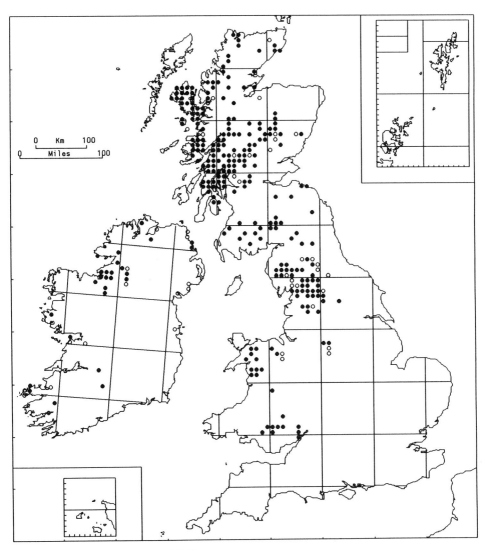

60/1. **Cololejeunea calcarea** (Lib.) Schiffn.

Locally frequent growing amongst other bryophytes or as pure patches on moist, shaded, steep, often vertical, basic rock-faces in wooded and treeless gorges and ravines, often near waterfalls, with *Lejeunea cavifolia*, *Marchesinia mackaii*, *Ctenidium molluscum* and *Neckera crispa*. Also occurs in crevices and under overhangs on damp, sheltered limestone faces and on damp N.- or E.-facing montane cliffs and in gullies, with *Ctenidium molluscum*, *Fissidens cristatus* and *Tortella tortuosa*. Commonest on limestone or other calcareous rocks (schists, basalts), also on siliceous rocks where there is some spray or intermittent seepage of calcareous water. Very rare as an epiphyte on ash or hazel in low-lying wooded ravines. Mainly a lowland species, it ascends to 825 m in the Scottish Highlands. GB 235+30*, IR 26+15*.

Monoecious; perianths and capsules common.

Widely scattered throughout Europe from the Balkans and Caucasus to Norway, Sweden, Faeroes. Also Azores, Canaries, Madeira, S.W. Asia.

H. J. B. Birks

265

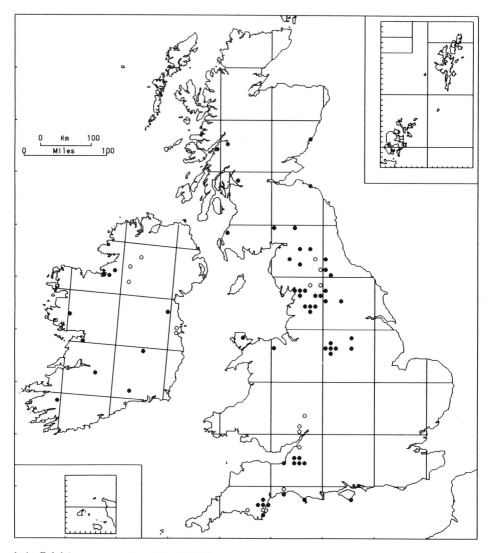

60/2. **Cololejeunea rossettiana** (Mass.) Schiffn.

On shaded basic rocks, often limestone, in gorges, streams and woods, and amongst other bryophytes or as pure patches on sheltered limestone rocks, cliffs and caves. It typically grows on dry rock in humid situations, with associates such as *Marchesinia mackaii*, *Neckera crispa*, *N. complanata* and *Tortella tortuosa*. Occasionally it is epiphytic on other bryophytes, e.g. *Thamnobryum alopecurum*. 0–365 m. GB 51+11*, IR 10+5*.

Monoecious; perianths and capsules common.

Scattered through the Mediterranean and S. Europe, extending up the west coast to W. France and the British Isles. Azores, Canaries, Madeira and N. Africa.

H. J. B. Birks

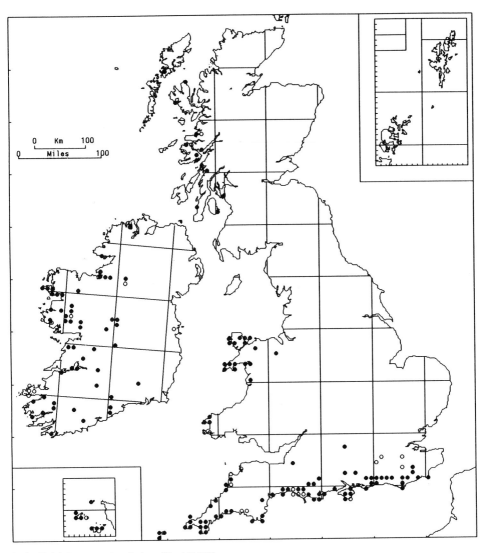

60/3. **Cololejeunea minutissima** (Sm.) Schiffn.

Epiphytic on trunks, lower branches, and upper twigs of willow, poplar, hazel, elder, ash, holly, oak, beech, sycamore, elm, blackthorn, and ivy in wind-exposed coastal scrub and in woods near the sea; also on gorse and tall *Calluna* stems, often on steep slopes near the sea, and on conifer trunks in the Isles of Scilly. Occurs more rarely on shaded mildly basic rocks and in ravines near the sea. Common epiphytic associates include *Frullania dilatata*, *Metzgeria fruticulosa*, *M. furcata* and *Ulota phyllantha*. Away from the coast, it tends to be replaced by *Lejeunea ulicina*. 0–150 m in Britain, extending to 250 m in Ireland. GB 99+18*, IR 51+7*.

Monoecious; perianths, capsules and gemmae common.

Mediterranean and S. Europe. Azores, Canaries, Madeira, Africa, E. and C. Asia, Caribbean, and N. America.

Its northernmost known world localities are in the Hebrides.

H. J. B. Birks

267

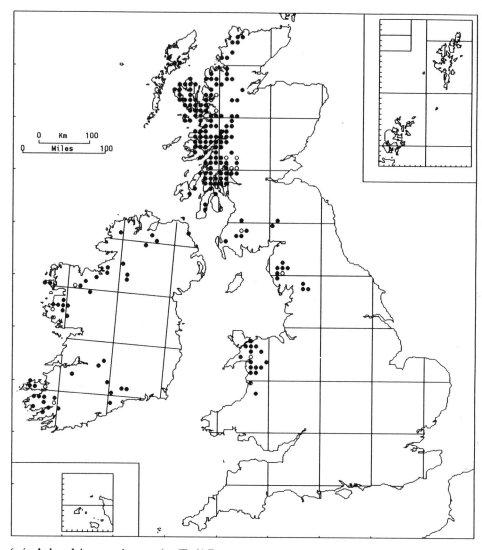

61/1. **Aphanolejeunea microscopica** (Tayl.) Evans

It grows in dense pure patches or as scattered stems amidst other bryophytes (*Frullania* spp., other Lejeuneaceae, *Metzgeria* spp.) on boulders, cliffs and rock-walls in shaded and sheltered situations such as low-lying wooded ravines, sea-cliffs, N.- or E.-facing stable screes, and the proximity of (and sometimes in) streams and waterfalls; also epiphytic on birch with *Frullania* spp., *Leptoscyphus cuneifolius*, *Plagiochila exigua* and *P. punctata*, more rarely on ash and hazel. Substrates are acid or mildly basic. When growing on rock, it tends to occur in damper, more shaded situations than *Harpalejeunea ovata*, and is less base-demanding than *H. ovata* or *Drepanolejeunea hamatifolia*. Mainly a lowland plant (0–300 m), extending to 520 m in W. Scotland and 610 m in S.W. Ireland. GB 163+14*, IR 43+4*.

Paroecious; perianths and gemmae uncommon, fruit rare.

Faeroes, Belgium, Luxemburg, Spain. Azores, Madeira, S.E. Asia.

H. J. B. BIRKS

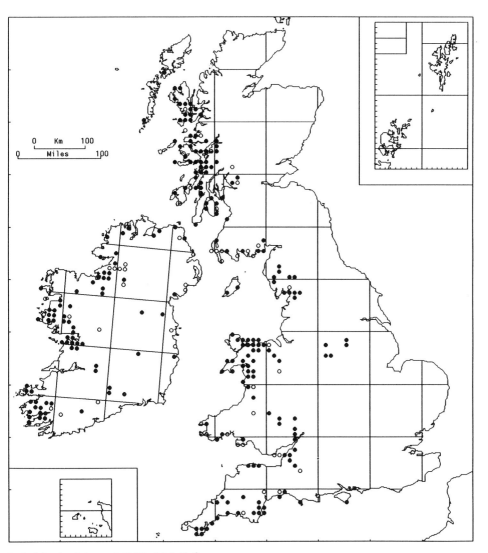

62/1. Marchesinia mackaii (Hook.) S. F. Gray

Locally frequent on dry, shaded or exposed basic, often highly calcareous rocks and vertical rock-walls in low-lying wooded or treeless gorges or ravines and on shaded rock outcrops, limestone pavements, and cliffs, especially in deciduous woodland with *Cololejeunea calcarea*, *Lejeunea cavifolia* and *Metzgeria furcata*. More rarely on shaded sea-cliffs and sheltered rocks near the sea and, very rarely, epiphytic on lower trunks of hazel, yew, hawthorn and elder. Confined to limestone or other basic rocks, occasionally on siliceous rocks but only where there is some intermittent calcareous seepage. Most localities, especially in Scotland, are within 2 km of the coast. 0–250 m in Britain, extending to 460 m in W. Ireland. GB 157+36*, IR 72+19*.

Monoecious; male branches, perianths and capsules common.

France, Spain, Portugal, Italy, Yugoslavia. Azores, Canaries, Madeira.

The northernmost known world localities are in the Hebrides.

H. J. B. Birks

269

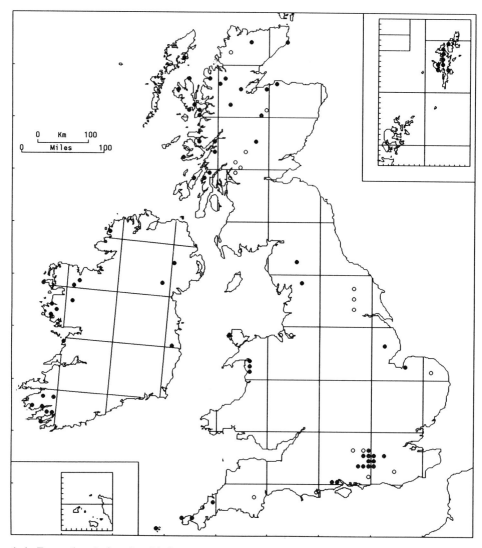

63/1. **Fossombronia foveolata** Lindb.

On damp, acidic, sparsely vegetated soil, usually at the edge of ponds, lakes or streams or in dried-up pools and ditches. It often grows in sites where sandy or gravelly soil is mixed with a peaty component but it is sometimes found on purer sands, gravels or peats, and on mud. *Callitriche* sp., *Isolepis setacea*, *Montia fontana*, *Haplomitrium hookeri*, *Riccardia incurvata*, *Riccia sorocarpa* and *Archidium alternifolium* have been noted as associates on Skye. It has also been recorded from china-clay workings in Cornwall and on clay in woodland rides in Surrey. Lakeside populations fluctuate in numbers from year to year, plants being most abundant in dry summers when the water-level is low and much suitable habitat is exposed. Lowland. GB 58+21*, IR 19.

Monoecious. Sporophytes common, recorded as early as May and as late as November but usually found between July and September.

Widespread in C. and N. Europe. N. America.

C. D. PRESTON

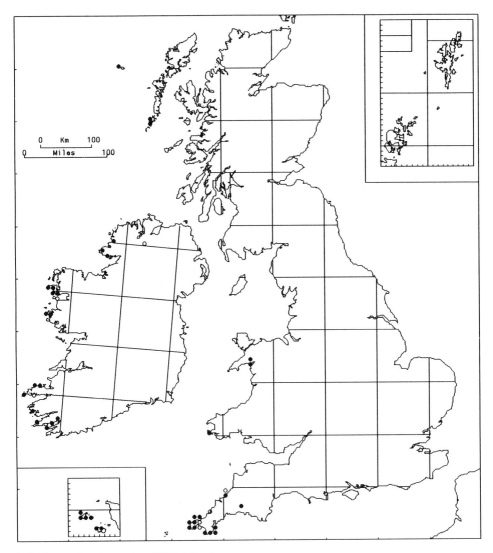

63/2. **Fossombronia angulosa** (Dicks.) Raddi

On moist cliffs, earth on cliff ledges, earthy pockets between rocks, and moist and flushed stream-banks and hedge-banks. Vegetatively the most vigorous of the British *Fossombronia* species, sometimes forming large, pure sheets on sea cliffs. It usually grows in rather shaded, sheltered sites and at its northernmost localities, in Barra and St Kilda, it is found on the sides of coastal ravines. Lowland. GB 18+4*, IR 17+3*.

Dioecious. Sporophytes frequent, recorded from January to June and rarely in other months.

A southern European species, recorded from all Mediterranean countries (including Turkey), reaching its northern limit in the British Isles. N. Africa, Macaronesia, south-eastern U.S.A., Caribbean.

C. D. PRESTON

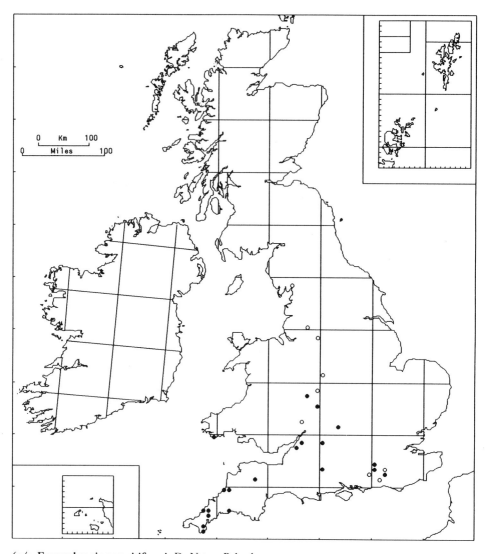

63/3. **Fossombronia caespitiformis** De Not. ex Rabenh.

A plant of damp soil in stubble-fields, pastures and woodland rides, often on acid clay. Also recorded from a sandy bank in Cornwall and a railway-cutting in W. Gloucestershire. Lowland. GB 20+9*.

Monoecious. Sporophytes frequent, recorded from July to April. Most records of the species are in the autumn (September–November) and this presumably reflects the main period of fruiting.

One of the commoner *Fossombronia* species in the Mediterranean region, extending northwards in W. Europe to the British Isles. N. Africa, Macaronesia.

The scarcity of this species in England, not easily explained in terms of its habitat requirements, perhaps reflects the fact that it is at its climatic limit. Its distribution in the British Isles is in need of critical revision.

C. D. Preston

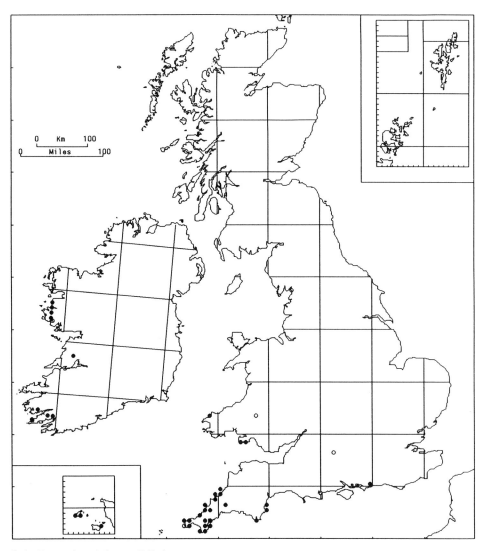

63/4. Fossombronia husnotii Corb.

Most characteristically grows over thin layers of soil on coastal cliff-ledges, rocky slopes and cliff-top paths, but also found on damp tracks, roadside banks, and clay soil on woodland rides. Some of its habitats, such as S.-facing cliffs in Cornwall and the Channel Islands, are subject to severe summer drought, during which the gametophytes die down to tuberous stems. These stems also allow the plants to withstand a certain degree of trampling on coastal paths. Lowland. GB 25+4*, IR 9.

Monoecious. Sporophytes frequent, developing in the winter months (September–April) in Cornwall, but also recorded in summer in Ireland.

A Mediterranean-Atlantic species, reaching its northern limit in the British Isles. N. Africa, Macaronesia.

C. D. Preston

273

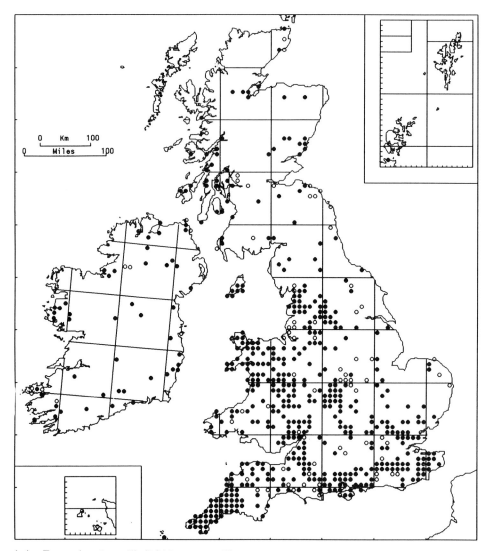

63/5a. **Fossombronia pusilla** (L.) Nees var. **pusilla**

A pioneer species of disturbed, damp, acidic soil in a wide variety of habitats including roadsides and tracksides, ditch-, stream- and river-banks, marshy pastures and stubble-fields. It is a characteristic member of the bryophyte flora of disturbed woodland rides. It shows some preference for loams and clays as opposed to peaty and sandy substrates and as a decided calcifuge is absent from soils derived from chalk and limestone. 0–450 m (Snowdon). GB 475+74*, IR 46+10*.

Monoecious. Sporophytes common, produced throughout the year in areas of heavier rainfall but ripening from autumn to spring in drier districts.

W., C. and S. Europe (including Turkey). N. Africa, Macaronesia, N. America.

The map shows all records of the species except those known to be the rare var. *maritima*.

C. D. PRESTON

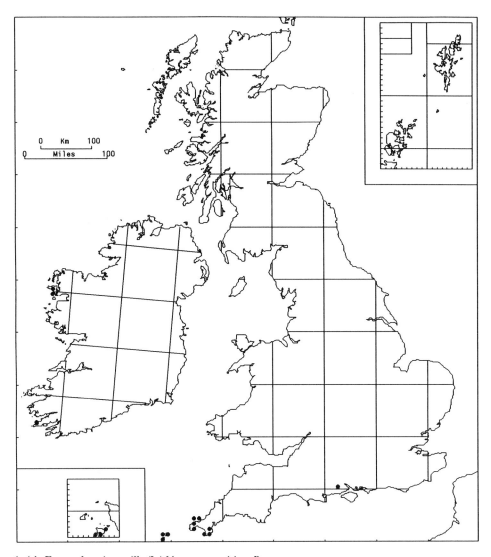

63/5b. **Fossombronia pusilla** (L.) Nees var. **maritima** Paton

On peaty soil or peaty sand on heaths and heathy cliffs, the floors of disused quarries, in non-calcareous dune-slacks, by pools, in shallow depressions, on tracks and in ruts. It is a perennial, usually occurring in more permanent, less disturbed sites than var. *pusilla*. It tolerates periods of submersion in winter-flooded habitats and of desiccation in localities subject to summer drought, surviving the latter as tuberous stems. Associated species include *F. husnotii*, *Cephaloziella* spp., *Archidium alternifolium*, *Bryum alpinum*, *Campylopus* spp., *Hypnum cupressiforme*, *Polytrichum juniperinum* and *P. piliferum*. Lowland. GB 11, IR 3.

Monoecious. Sporophytes frequent, recorded from December to June.

Described, new to science, in 1973 (Paton, 1973a). Outside the British Isles known from the Mediterranean coast of France and recently reported from Australia; the relationship of var. *maritima* to forms of *F. pusilla* elsewhere in the Mediterranean region requires further study.

C. D. PRESTON

275

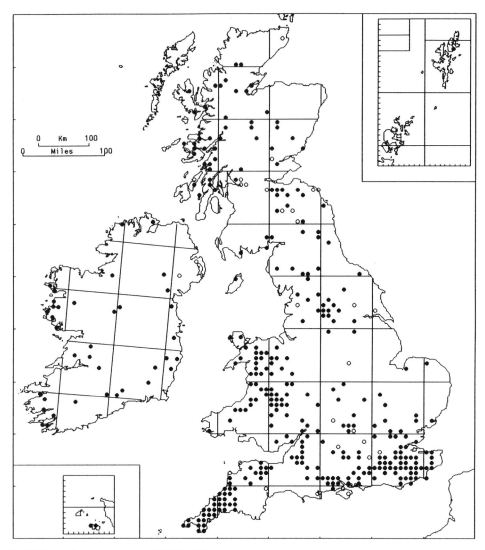

63/6. **Fossombronia wondraczekii** (Corda) Dum.

On damp acidic soil on disturbed woodland rides, tracksides, stream-sides, ditch banks, marshy pastures, stubble-fields and at the edge of ponds, lakes and reservoirs. In the chestnut-woods of Kent it behaves as a 'coppice plant', tending to occur on recently disturbed soil in the year after coppicing. Although very similar ecologically to *F. pusilla* and often growing with it, *F. wondraczekii* extends on to more acid, less fertile soils, occurs by lakes and reservoirs – a habitat not exploited by *F. pusilla* – and although rarer in England tends to be commoner than *F. pusilla* in many areas of Scotland. 0–485 m (The Storr, Skye). GB 306+30*, IR 37+2*.

Monoecious. Sporophytes common, recorded throughout the year but perhaps most frequently in autumn. Throughout mainland Europe. N. Africa, eastern N. America.

C. D. PRESTON

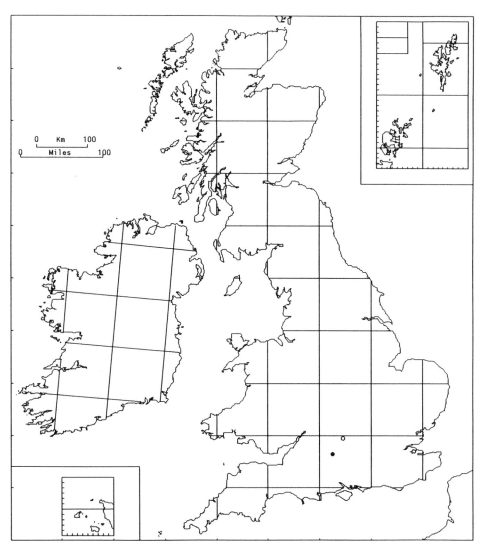

63/7. **Fossombronia crozalsii** Corb.

In woodland rides. In Wiltshire it has been found on moist sandy loam in a wide, rutted grassy ride of a secondary wood which has been converted to a conifer plantation. *F. pusilla, F. wondraczekii, Jungermannia gracillima, J. hyalina, Riccia glauca, R. sorocarpa, R. subbifurca, Scapania irrigua* and *Pseudephemerum nitidum* grew in the immediate vicinity (Paton, 1973a). In Berkshire, it was found on a ride in a larch plantation on moderately acid sandy soil. Lowland. GB 1 + 1*.

Monoecious. Sporophytes ripe August and September.

In Europe known only from Britain and France (the type was collected under *Cistus* and *Erica* scrub in S. France); elsewhere recorded from N. Africa and the Canaries. Perhaps conspecific with the commonest W. African species of *Fossombronia, F. occidento-africana* (Jones & Harrington, 1983).

Discovered 1918 in Chisbury Wood near Great Bedwyn, Wiltshire, where it was refound by J. A. Paton in 1972. In Berkshire, it was collected by E. W. Jones in 1938 but has never been rediscovered.

C. D. PRESTON

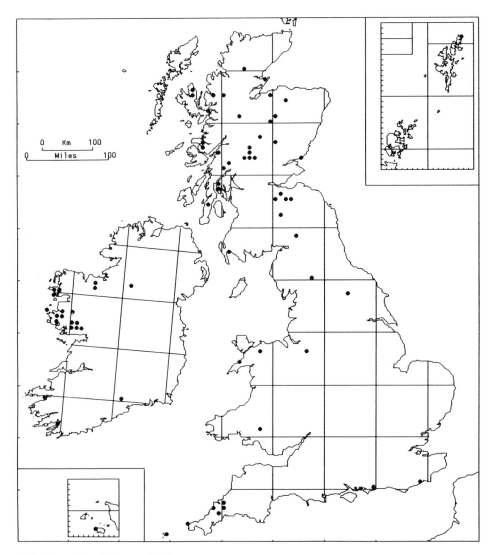

63/8. Fossombronia incurva Lindb.

Found as scattered plants or in small patches on sparsely vegetated soil or in low bryophyte turf in a wide range of habitats where competition is low, including roadsides, lay-bys, tracks and paths, disused railway lines, the edges of lakes and reservoirs, streamsides, dune-slacks, gravel-pits and the floors of disused quarries. Most often found on moist or flushed, sandy or gravelly soil, but also recorded from clay, china clay and peaty soil. Associates include *F. fimbriata*, *F. wondraczekii*, *Haplomitrium hookeri*, *Riccardia incurvata*, *Archidium alternifolium* and bulbiliferous species of *Pohlia*. 0–500 m (Lochan na Lairige). GB 50, IR 20.

Dioecious. Sporophytes frequent, recorded from March to October.

Endemic to Europe; limited to a rather compact area bordered by the British Isles, Germany, Poland and S. Scandinavia.

Overlooked by British bryologists until 1964, when it was found at Lochan na Lairige by A. C. Crundwell. It doubtless awaits discovery at many localities in addition to those shown on the map.

C. D. Preston

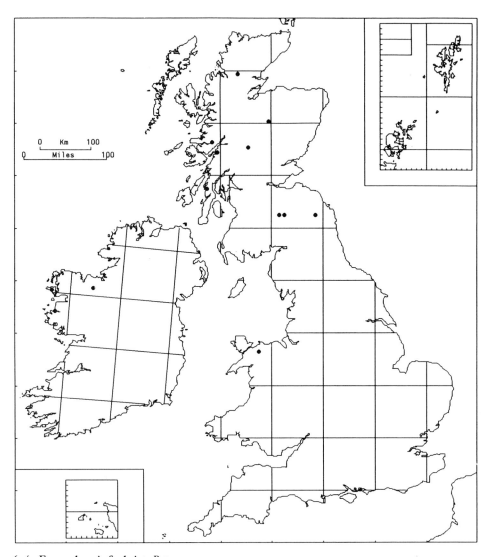

63/9. Fossombronia fimbriata Paton

Occurs as scattered plants on damp gravelly, sandy or schistose soil on streamsides, lake margins, roadsides, paths and the floor of a disused quarry. Characteristic associates include *F. incurva*, *F. wondraczekii*, *Haplomitrium hookeri*, *Riccardia incurvata* and bulbiliferous *Pohlia* species. 0–350 m. GB 10, IR 3.

Dioecious. Gametophytes have been recorded from June to September; in unfavourable seasons they perennate as tuberous stems. They often bear sex organs but mature sporophytes have not been observed in the wild, although they developed in cultivation on plants collected in August.

F. fimbriata was discovered in Westerness in 1967 and described, new to science, in 1974 (Paton, 1974a). Although unmistakable in appearance it is easily overlooked and is almost certainly more frequent than the map suggests. It has not yet been found on the European mainland.

C. D. PRESTON

279

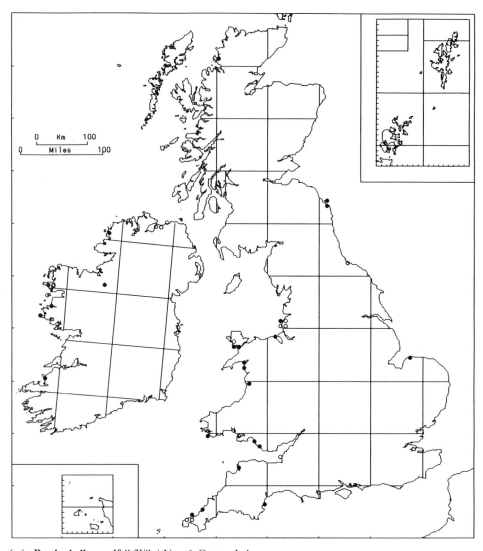

64/1. **Petalophyllum ralfsii** (Wils.) Nees & Gott. ex Lehm.

A coastal species of calcareous sand-dunes where it may be locally frequent in and along the margins of slacks. Most conspicuous in winter and spring, it disappears almost completely during periods of drought. Frequent associates include *Aneura pinguis*, *Leiocolea badensis*, *Preissia quadrata*, *Riccardia chamedryfolia*, *R. incurvata*, *Barbula* spp., *Bryum* spp., *Dicranella varia* and *Trichostomum* spp. Lowland. GB 18+11*, IR 6+8*.

Dioecious; sporophytes frequent from December to June. Perennation through periods of drought is by means of tubers packed with lipid.

Widespread in the Mediterranean region including N. Africa and Turkey, extending northwards along the Atlantic seaboard to Britain. Southern U.S.A.

J. G. Duckett

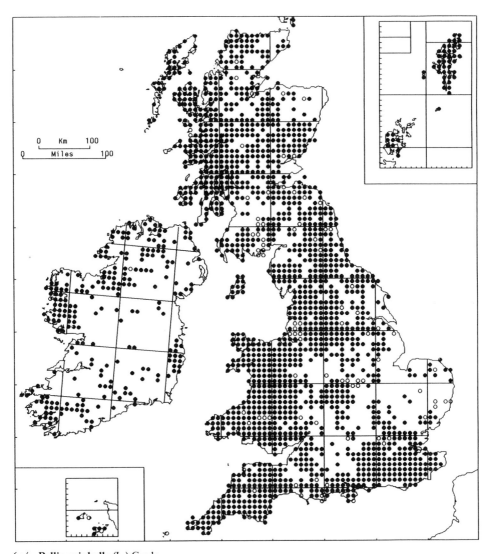

65/1. **Pellia epiphylla** (L.) Corda

A calcifuge forming extensive sheets on wet soil and acidic rock outcrops, often in the presence of, but not directly on, peat. It occurs on the banks of streams, rivers and ditches, but is also common on the ground in woods and in the open. 0–1230 m (Einich Cairn). GB 1654+86*, IR 237+2*.

Monoecious and nearly always fertile; sporophytes common, maturing in spring. Without special means of asexual reproduction.

Widespread in temperate Europe, Asia and N. America, south to N. African mountains, Himalaya and Texas.

M. E. NEWTON

281

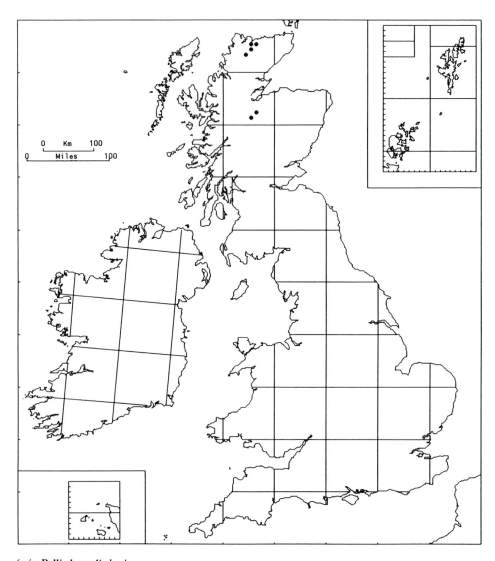

65/2. Pellia borealis Lorbeer

Apparently confined to the peaty margins of lochs, generally at low altitudes. 60–390 m (Tomanraid). GB 6.
Monoecious; sporophytes maturing in early summer. Without special means of asexual reproduction.
Elsewhere, the species is recorded from Belgium, Germany, Poland and the U.S.S.R.

First discovered in Britain in 1984 (Loch Meadie), *P. borealis* has been overlooked as *P. epiphylla*, from which it is morphologically indistinguishable. There is strong chromosomal evidence (Newton, 1986) to suggest that the two are genetically isolated and have been so for a considerable period. A simple cytological technique allows for reliable identification.

M. E. NEWTON

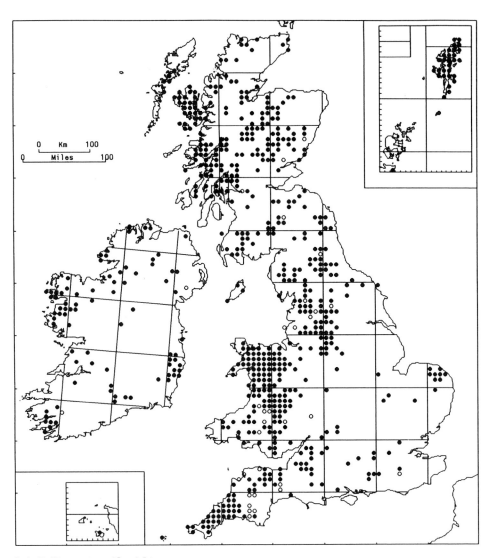

65/3. **Pellia neesiana** (Gott.) Limpr.

On soil in non-calcareous marshes and wet flushes. In lowland England, it commonly grows on damp clay in woodland rides. Its pH preference is for neutral conditions, intermediate between *P. epiphylla* and *P. endiviifolia*. 0–960 m (Ben Lawers). GB 557+30*, IR 82+2*.

Dioecious and usually fertile. Sporophytes are frequent only where the plant occurs on soil. Special means of asexual reproduction are absent.

Boreal-montane, with a more northerly distribution than *P. epiphylla*, reaching Iceland and Greenland, and commonly above the tree-limit in the Scandinavian mountains. Asia, N. America.

This species has been much confused with *P. epiphylla*. Supposed differences in coloration are unreliable, and identification should not be attempted in the absence of female inflorescences. Since colonies of *P. neesiana* are frequently unisexual, this has led to considerable under-recording.

M. E. NEWTON

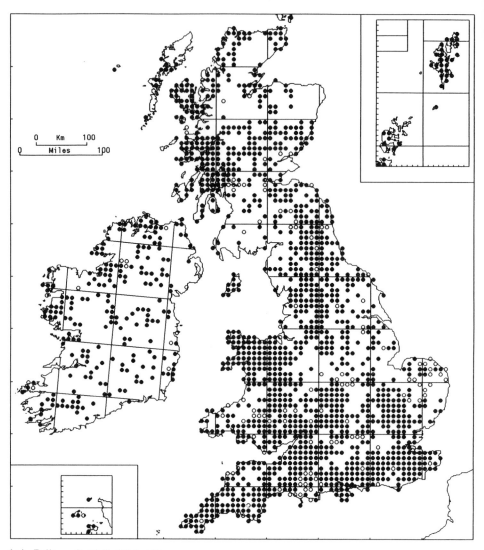

65/4. **Pellia endiviifolia** (Dicks.) Dum.

Calcicolous, on a wide range of wet substrates including sand, soil, outcropping rocks, and, occasionally, peat. It is often a conspicuous part of the ground flora in woodland and is common on stable dunes, in soil-filled rock crevices and in and by running water, where it sometimes forms tufa. Descending almost to the level of high tide in certain coastal regions, it is generally distributed over the lowlands and has been recorded at altitudes up to 730 m (Stob Ghabhar). GB 1365+114*, IR 213+12*.

Dioecious and usually fertile; sporophytes frequent, maturing in spring. Vegetative propagation is achieved by the development of repeatedly bifurcating proliferations at thallus tips in autumn and winter.

Widely distributed in Europe, Mediterranean coastal regions of N. Africa, Asia and western N. America.

Plants referred to this species are morphologically (Schuster, 1981) and cytologically (Newton, 1988) diverse, requiring taxonomic revision.

M. E. NEWTON

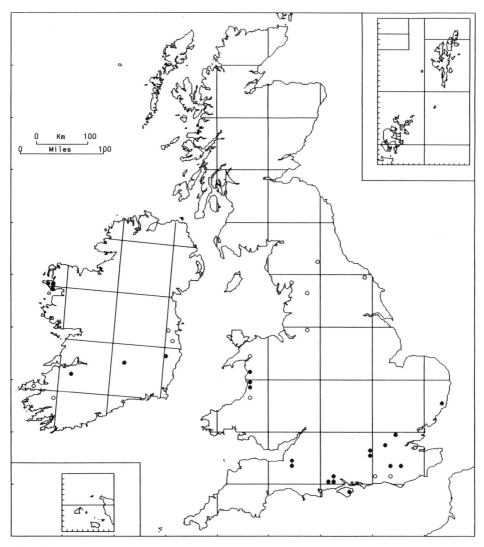

66/1. **Pallavicinia lyellii** (Hook.) Carruth.

A species of moist shaded sandstone rocks, associated with *Calypogeia* spp., *Conocephalum conicum*, *Kurzia* spp., *Pellia epiphylla* and *Tetraphis pellucida*. It also grows on the sides of wet, shaded ditches in bogs. Lowland. GB 16+9*, IR 6+5*.

Dioecious; sporophytes rare. In some localities only one sex is present. Specialized methods of vegetative dispersal are absent.

Widely distributed in Europe from S. Scandinavia south to the Azores. C. and E. Asia, N. America, Oceania, S. America, New Zealand.

Notably rarer in Britain than in other parts of its world range. In eastern N. America it is a common plant of wet hollows in woodlands.

J. G. DUCKETT

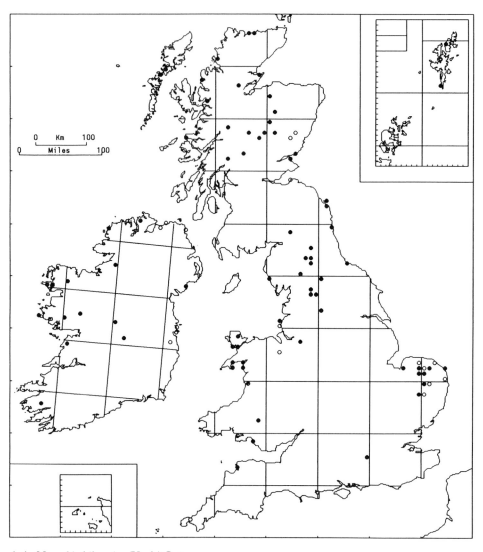

67/1. **Moerckia hibernica** (Hook.) Gott.

A plant of calcareous dune-slacks where associated species may include *Petalophyllum ralfsii*, *Amblyodon dealbatus* and *Meesia uliginosa*. Rarely occurring inland in base-rich flushes, fens, quarries, on ditch-banks and on silty or marshy ground. Sand-dune plants appear drought-tolerant, new shoots innovating from the midrib of plants surviving the summer. Mainly at low altitudes but rarely ascending to 750 m. GB 61+15*, IR 14+6*.
　　Dioecious; sporophytes rare, spring and early summer.
　　Europe, especially C. and W., Faeroes. Siberia (Yenisei), northern N. America.

A. J. E. Smith

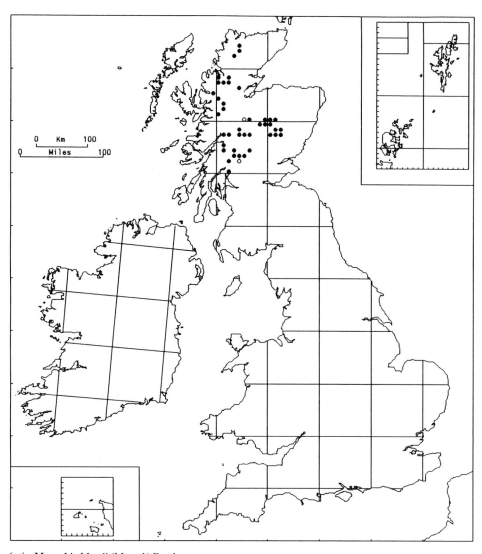

67/2. **Moerckia blyttii** (Moerck) Brockm.

A characteristic soil species in areas of late snow-lie but also occurring in short turf and scree. Habitats include peaty soil and turfy banks in boggy hollows, streamsides, margins of pools and lochs, and sheltered gullies where snow accumulates. Associated bryophytes include *Anthelia juratzkana*, *Barbilophozia floerkei*, *Cephalozia bicuspidata*, *Nardia scalaris*, *Kiaeria starkei* and *Oligotrichum hercynicum*. 750 m (Beinn Eighe) to 1190 m (Cairngorms). GB 41+2*.

Dioecious; capsules occasional, July, August.

Montane and subarctic Europe, Faeroes, Iceland, Svalbard. Novaya Zemlya, Caucasus, western N. America, Newfoundland, S. Greenland.

A. J. E. SMITH

287

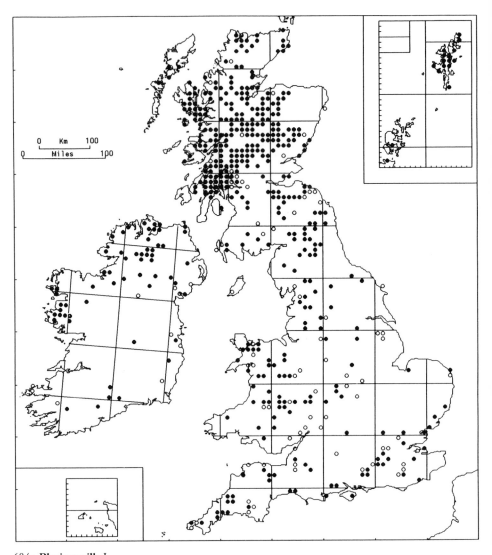

68/1. **Blasia pusilla** L.

On moist non-calcareous soil on waste ground, in old quarries and clay-pits, on damp tracks and roadsides, in ditches and, occasionally, arable fields. Less common in natural habitats, chiefly amongst gravel by streams and rivers, and on lake shores. 0–1000 m (Glas Tulaichean). GB 404+75*, IR 58+10*.

Dioecious. Capsules rare, ripe in spring. Stellate gemmae, or gemma receptacles, or both, almost always present.

Circumboreal. Throughout Europe but rare in south.

A. C. CRUNDWELL

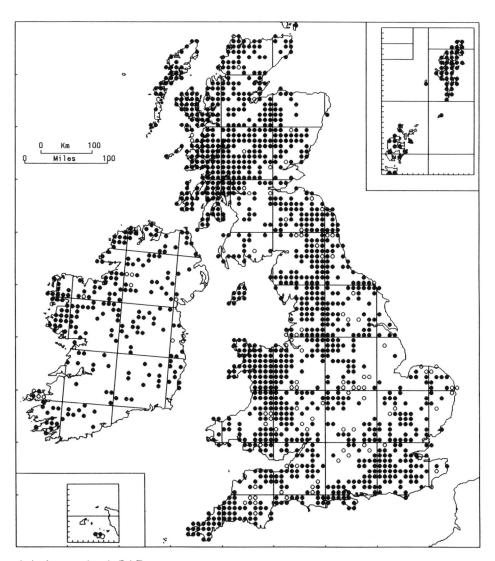

69/1. Aneura pinguis (L.) Dum.

A species found in a variety of mire communities, generally creeping among other bryophytes. It is most characteristic of mesotrophic to eutrophic conditions but is by no means restricted to them and also occurs (although usually in small quantity) in base-poor situations such as seepage-lines in acid bogs. Associates range from *Calliergon cuspidatum*, *Campylium stellatum* and *Drepanocladus revolvens* in fens and calcareous flushes to *Sphagnum* spp. under more acidic conditions. It also occurs in various other damp or wet habitats, which are generally at least mildly basic, including damp banks and turf, flushed rocks, lake and stream margins (where it sometimes grows submerged) and dune-slacks. 0–1000 m (Killin). GB 1275+94*, IR 212+14*.

Dioecious; capsules rare, noted as maturing at various times of year. Gemmae absent.

Circumboreal, extending south to N. Africa, Macaronesia and northern S. America. Widespread in Europe but rare in the Mediterranean region.

M. M. YEO

289

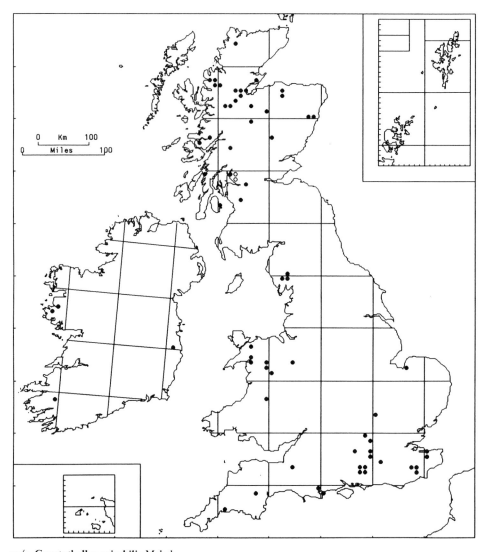

70/1. **Cryptothallus mirabilis** Malmb.

A parasitic liverwort of wet acid peat, growing a few centimetres below the soil surface. Its commonest habitat is under *Sphagnum* (especially *S. fimbriatum*, *S. palustre* and *S. recurvum*) in boggy birch-woods, but it also grows between the tussocks in wet *Molinia* communities, and has been found under *Pellia*. Lowland. GB 62+2*, IR 4.

Dioecious. Capsules common. Gemmae unknown.

In Scandinavia, Russia and N. America.

As this liverwort grows underground and has to be burrowed for, its distribution, both in the British Isles and elsewhere, is certainly very imperfectly known.

A. C. CRUNDWELL

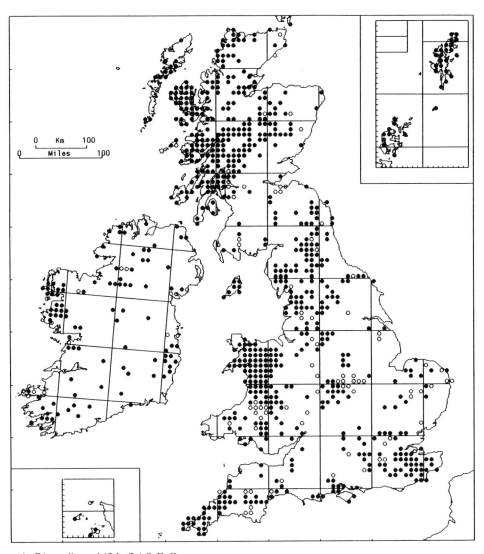

71/1. Riccardia multifida (L.) S. F. Gray

This species grows in a variety of mire communities, from acidic bogs to strongly calcareous fens, but is perhaps most characteristic of peaty, mildly basic flushes with associates such as *Campylium stellatum*. It is also frequent on flushed rocks, especially in sheltered gullies. Other habitats include damp banks, grassland and tracks, wet soil in woods, and lake margins. Ecological distinctions between this species and *R. chamedryfolia* are not very clear but *R. multifida* generally appears to be more restricted to permanently moist conditions. 0–800 m (Ben Lawers). GB 677+86*, IR 97+15*.

Autoecious; capsules frequent. Gemmae rather rare.

Circumboreal. Widespread in Europe but rare in the Mediterranean region.

M. M. YEO

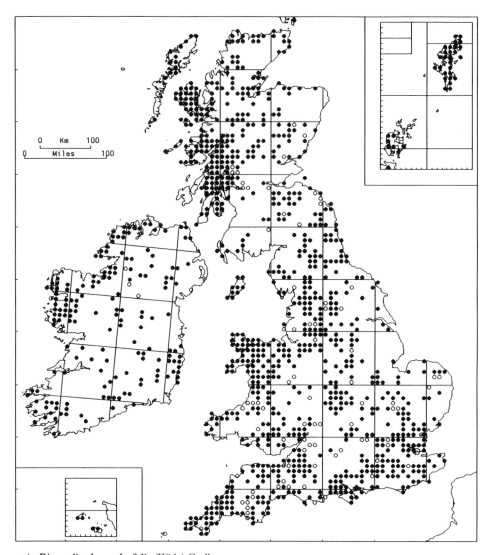

71/2. **Riccardia chamedryfolia** (With.) Grolle

A species of damp or wet habitats, avoiding strongly acidic conditions. It is widespread in topogenous and soligenous mires and on dripping rocks and wet ledges, particularly in stream gullies where it can tolerate considerable shade. It is also found on damp soil in other situations, including banks, tracks, woodland rides, pond and lake margins and dune-slacks, often on mineral soils such as clay. It can grow in habitats liable to dry out at certain times of year as, for example, on chalky banks with *Leiocolea turbinata*. 0–950 m (Beinn Dorain). GB 880+98*, IR 168+6*.

Autoecious; capsules occasional to frequent, maturing at various times of the year. Gemmae occasional.

Circumboreal. Widespread in Europe but rare in the south.

In the earlier years of the BBS mapping scheme, *R. chamedryfolia* and *R. multifida* were liable to confusion (Little, 1968). The maps must therefore contain errors, although these are unlikely to affect the overall patterns.

M. M. YEO

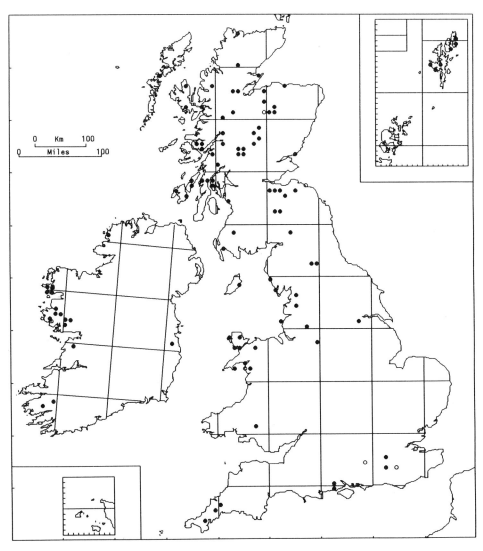

71/3. **Riccardia incurvata** Lindb.

A species of moist gravelly or sandy soil in open habitats that are generally only sparsely colonized by higher plants, as, for example, on lake margins, beside streams, in quarries and sand-pits, in dune-slacks and on damp roadsides and tracks. The substrate may be either acidic or calcareous. As well as growing with various common bryophytes of damp ground it is also frequently associated with species with weedy tendencies such as *Fossombronia incurva*, bulbiliferous *Pohlia* spp. and, more rarely, *Haplomitrium hookeri*. 0–670 m (Schiehallion). GB 86+5*, IR 16.

Dioecious; capsules rare. Gemmae frequent.

Widespread in Europe but absent from the Mediterranean region. Elsewhere known only from N. America. Probably under-recorded because of its small size and similarity to *R. chamedryfolia*.

M. M. YEO

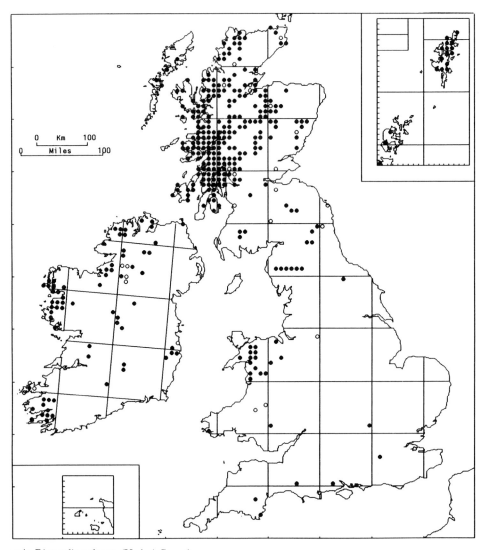

71/4. Riccardia palmata (Hedw.) Carruth.

Forms small patches, or occasionally more extensive sheets, on decorticated logs and stumps which have rotted to a soft spongy texture, most typically in sheltered woodland but also in open habitats such as on moorland. Frequent associates include *Cephalozia* spp., *Nowellia curvifolia* and *Scapania umbrosa*. In common with some other liverworts characteristic of rotting wood, such as *N. curvifolia*, in parts of Scotland and Ireland it also grows frequently on moist peaty soil on banks and among rocks. *R. latifrons* occasionally grows with it in these situations. 0–450 m (Creag Meagaidh). GB 281+18*, IR 68+8*.

Dioecious; capsules occasional, maturing in spring and summer. Asexual reproduction is by means of gemmae on branch apices.

Circumboreal. N., C., W. and E. Europe.

M. M. YEO

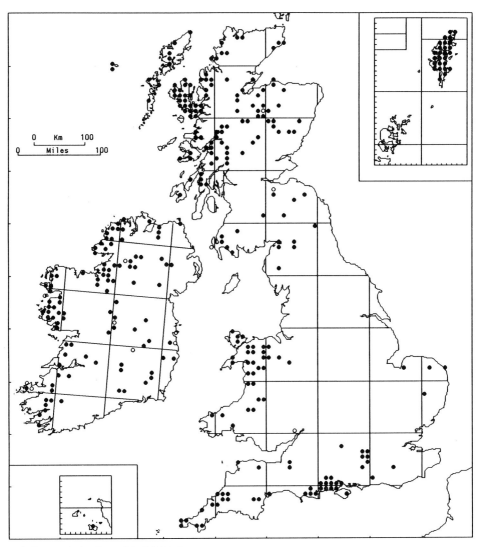

71/5. **Riccardia latifrons** (Lindb.) Lindb.

Usually found as scattered plants or small patches in acid mires, growing on damp peaty banks, sphagnum hummocks, tussocks of *Molinia* and decaying litter. In fens and other base-rich mires it is usually restricted to acidified tussocks. *Cephalozia* spp., *Kurzia pauciflora* and *Odontoschisma sphagni* are frequent associates. It is also found, less commonly, with species such as *Nowellia curvifolia* and *R. palmata* on decorticated logs in bogs and woodland. 0–880 m (Coire an Lochain). GB 231+5*, IR 98+8*.

Autoecious; capsules occasional, maturing in spring and summer. Asexual reproduction is by means of gemmae on branch apices.

Circumboreal. Widespread in Europe but rare in the south.

This species has been confused in the past with other *Riccardia* spp., in particular *R. chamedryfolia*. Old literature records are therefore unreliable.

M. M. YEO

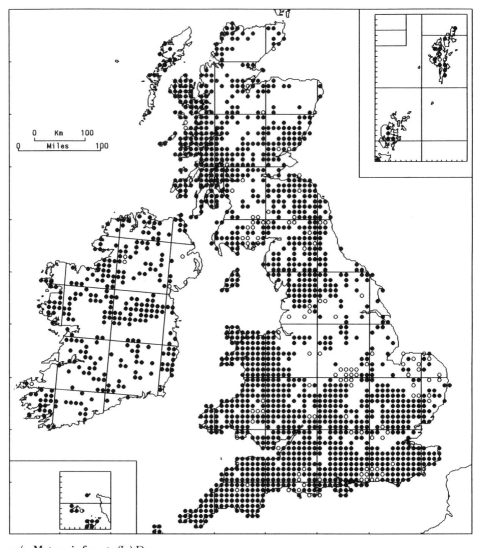

72/1. Metzgeria furcata (L.) Dum.

This species has a wide ecological range, occurring on all but the most acid trees and rocks and being tolerant of drought. However, it is seldom found on limestone or on decaying wood. There is a wide range of associates, depending on the substrate. It is a pioneer species on the middle part of tree-trunks (*c.* 2.5 m above ground-level) where *Hypnum cupressiforme* is the most frequent associate. Mainly lowland, to 600 m. GB 1504+125*, IR 283+6*.

Dioecious; female plants common, male plants less so; capsules occasional. Gemmae are frequently produced on the thallus margin.

All Europe but rare in the far north. Very widespread in the Northern Hemisphere, and occurring also in the Southern Hemisphere.

N. G. Hodgetts

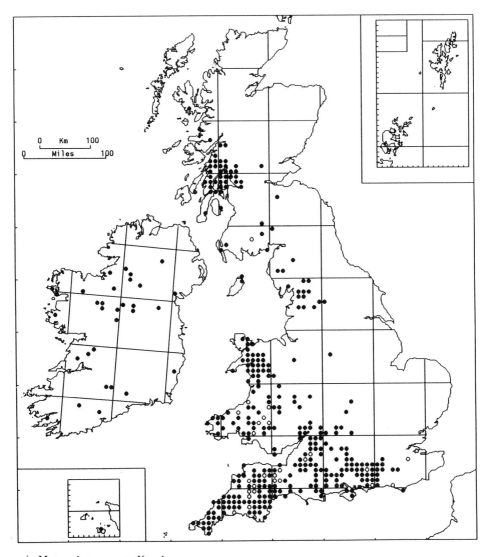

72/2. Metzgeria temperata Kuwah.

An epiphyte which favours more base-deficient bark than *M. fruticulosa*, being found chiefly on the trunks and lower branches of willow, beech, oak and elder. It occasionally grows on rock, particularly base-deficient rock such as granite. *Lejeunea ulicina* is a frequent associate. Mainly a lowland plant of sheltered, moist sites, but occasionally ascending to 350 m. GB 315+27*, IR 31.

Dioecious; fairly often fertile with female plants more frequent than male, sporophytes not found in the British Isles. Gemmae abundant.

W. Europe, extending north to Britain and east to Austria. Madeira, Japan, N. America.

N. G. HODGETTS

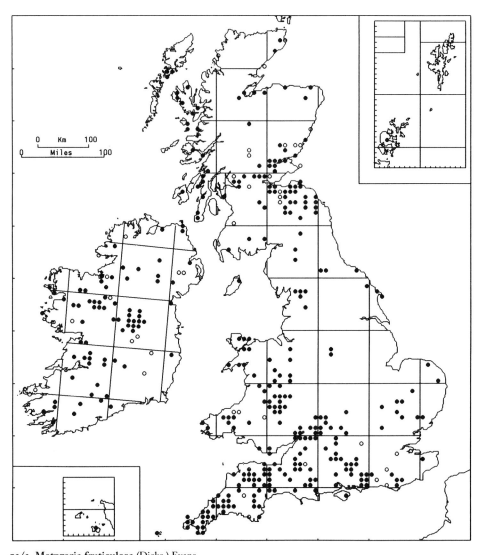

72/3. Metzgeria fruticulosa (Dicks.) Evans

An epiphyte found mainly on elder, but also on willow, ash and hazel, and less frequently on other trees. It avoids the more acid types of bark. It is often associated with *Frullania dilatata* and *Cryphaea heteromalla*, and sometimes with species of *Orthotrichum*, *Ulota* and *Zygodon*. It is usually a lowland species of sheltered sites, but has been found up to about 450 m. GB 279+29*, IR 73+14*.

Dioecious; female plants more frequent than male, sporophytes very rare. Gemmae abundant.

W. and C. Europe, extending north to S. Scandinavia and east to the Baltic region. Madeira, Mexico.

Metzgeria fruticulosa and *M. temperata* have only recently been separated (Paton, 1977a). Field and literature records made before *M. temperata* was recognized have had to be discarded, as have subsequent ambiguous records referred to *M. fruticulosa*. *M. fruticulosa* is consequently under-recorded.

N. G. Hodgetts

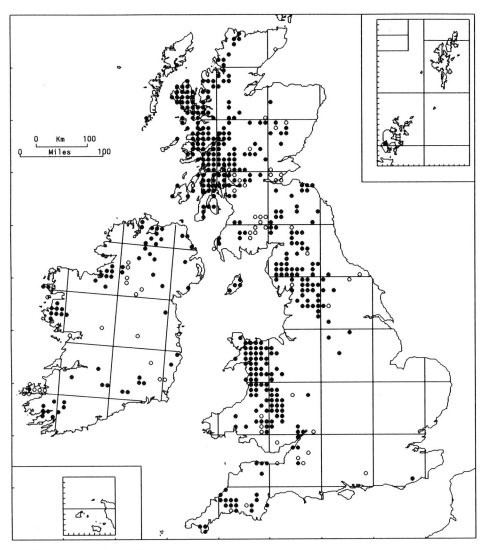

72/4. **Metzgeria conjugata** Lindb.

A species of moist, shady base-rich rocks, most of the records being from ravines, sheltered river valleys and cliffs. It grows in pure patches or creeping among mosses on a wide variety of rock types, but it also occurs, more rarely, on tree-trunks and roots, sometimes forming mixed patches with *M. furcata*. 0–800 m (Skye). GB 381+42*, IR 65+21*.

Autoecious, usually fertile; capsules frequent. Gemmae are absent.

Most of Europe except the north-east and far north. Macaronesia, southern Africa, N., C., and S. America.

N. G. Hodgetts

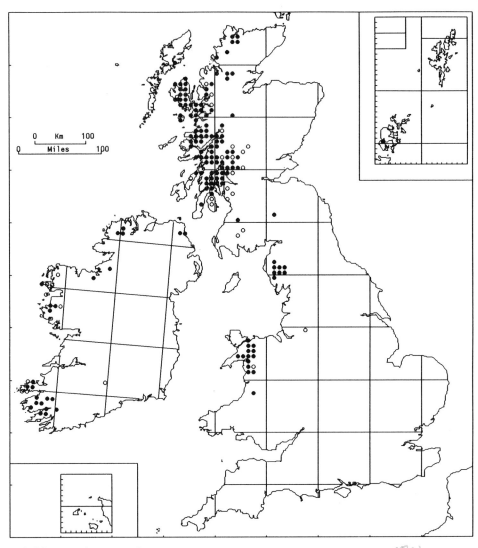

72/5. **Metzgeria leptoneura** Spruce

On wet acidic or basic rocks in sheltered situations. Records are mostly from shady, wooded ravines, where the plant often grows in the spray zone of waterfalls. More rarely it occurs on moist and shaded rocks where trees are lacking, for example above the tree-limit on mountains. It often grows on dripping-wet rocky banks over mosses such as *Eurhynchium* spp., *Hyocomium armoricum* and *Philonotis fontana*. 0–900 m (Argyll). GB 122+23*, IR 24+8*.

Dioecious; only female in the British Isles.

In Europe, an Atlantic species, confined to the Faeroes and the British Isles. Macaronesia, tropical and subtropical mountains in Asia, Africa and America, Australia, New Zealand.

N. G. HODGETTS

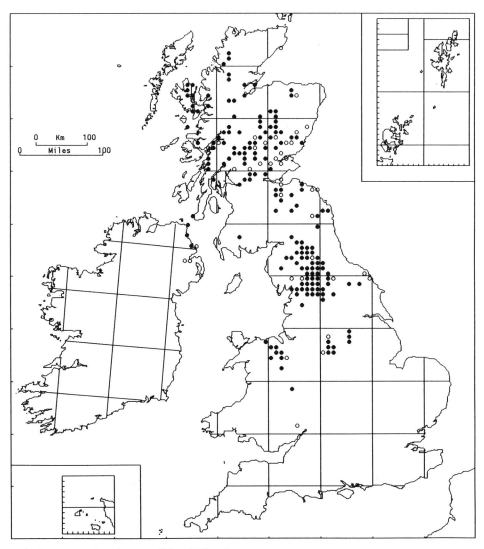

73/1. **Apometzgeria pubescens** (Schrank) Kuwah.

A strong calcicole of dry, shaded limestone rock, often on vertical surfaces. On Carboniferous limestone in England and Wales, the main habitat is ravines and N.-facing cliffs, where it may have a wide range of calcicolous associates. In Scotland, it sometimes grows on siliceous rocks such as sandstone and Old Red Sandstone conglomerate, but always where strongly base-rich. o–680 m (Creag an Dail Bheag, Cairngorms). GB 167+39*, IR 3+3*.

Dioecious; capsules unknown in Britain, no data on sexuality. It has no obvious means of vegetative propagation.

Mountains of E., C. and W. Europe; absent from the north except Scotland and W. Norway. Widespread in Asia and N. America.

M. O. HILL

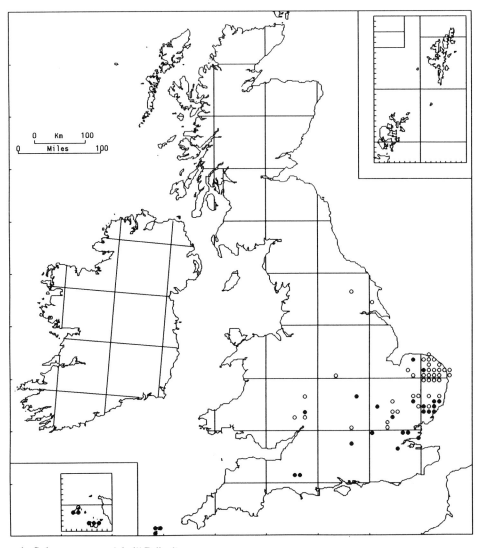

74/1. **Sphaerocarpos michelii** Bellardi

A weed of cultivated ground, recorded from non-calcareous clay, loam or sandy soil in neglected gardens, nurseries, fallow land, cereal stubble and in fields of turnips, cabbages, sugar-beet, rape and newly-sown clover. In the Isles of Scilly it is frequent in flower-fields, and in the Channel Islands it also grows on roadside banks, pathsides, waste ground and damp tracks. Associates include *Riccia glauca*, *R. sorocarpa*, *Bryum bicolor*, *B. rubens*, *Dicranella staphylina*, *Ditrichum cylindricum*, *Phascum cuspidatum* and *Pottia truncata*. Lowland. GB 26+40*.

Dioecious. Sporophytes abundant, maturing from late November to June. A single thallus can produce capsules over a period of six months (W. H. Burrell, unpublished letter). Spores are shed in tetrads.

S. and C. Europe, as far north as the British Isles. N. Africa, southern N. America, S. America.

Sphaerocarpos michelii might be a long-established introduction rather than a native species. It is probably decreasing because land is now usually ploughed and reseeded after harvest. The concentration of old records in Norfolk reflects the activities of W. H. Burrell in the early years of this century.

C. D. Preston

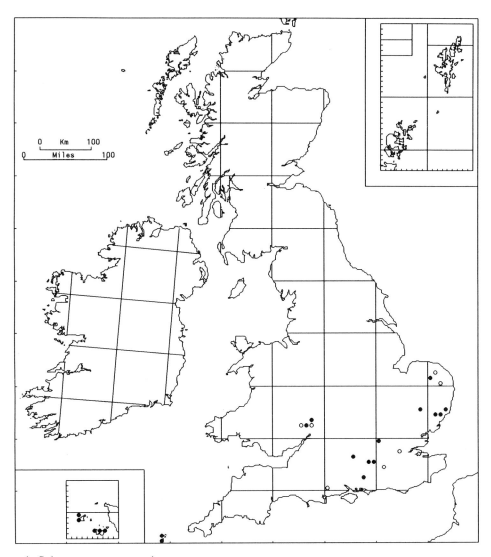

74/2. **Sphaerocarpos texanus** Aust.

A weed of cultivated ground, growing on non-calcareous loam or sandy soil in neglected gardens, nurseries, fallow land, cereal stubble and fields of cabbages, cauliflowers, rape, clover and (in the Isles of Scilly) flowers. In the Channel Islands it also occurs on hedge-banks, waste ground and the floor of a disused quarry. It is often accompanied by *S. michelii*, which has similar ecological requirements. Lowland. GB 14+7*.

Dioecious. Sporophytes abundant, maturing from December to June. Spores are shed in tetrads of two males and two females.

S. and C. Europe, reaching its northern limit in the British Isles. Otherwise known from N. Africa, southern N. America, S. America and, perhaps as an introduction, from Australia.

One might speculate that *S. texanus*, like *S. michelii*, is a long-established introduction rather than a native species in the British Isles.

C. D. PRESTON

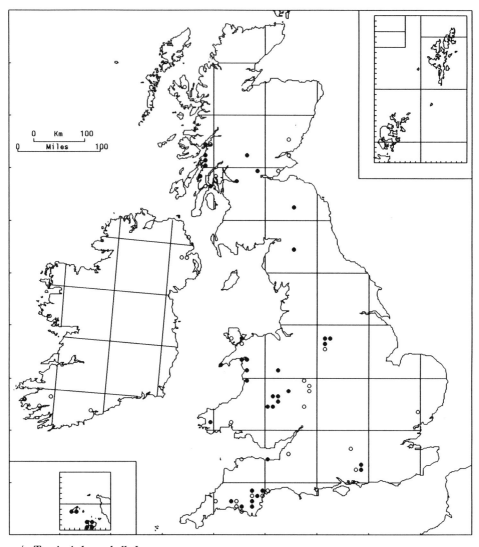

75/1. **Targionia hypophylla** L.

On earthy, rocky or sandy roadside banks, on thin layers of soil over walls and rocks and, rarely, in coastal turf. *Targionia* usually occurs in sites which dry out in summer: it is one of the most xerophytic of British liverworts. Although not a strict calcicole, it is often associated with limestone or basic igneous rock. Lowland. GB 38+27*, IR 7*.

Monoecious; sporophytes frequent, maturing in winter or early spring.

A southern species, frequent in the Mediterranean countries and reaching its northern limit in the British Isles. Widespread in warm regions of the world, south to S. Africa, Australia, S. America.

The genera *Targionia*, *Reboulia* and *Preissia* form an interesting series, with *Targionia* most and *Preissia* least adapted to existence in Mediterranean microclimates. This difference is reflected in their British distribution.

C. D. PRESTON

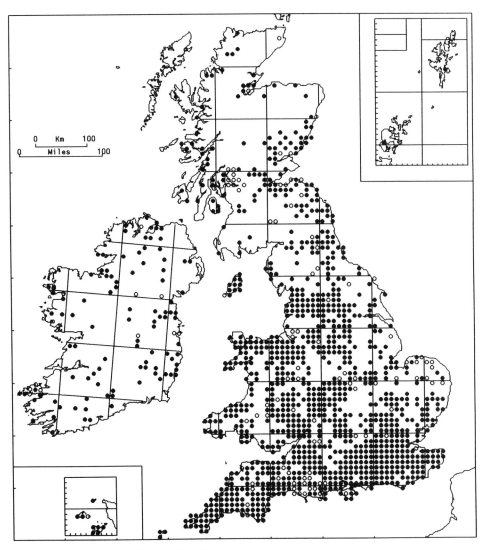

76/1. Lunularia cruciata (L.) Dum. ex Lindb.

A weed of paths, trampled soil, flower-pots and the damp bases of walls, often found in streets, gardens, churchyards and greenhouses. It also occurs in more natural communities, most characteristically on the banks of streams and rivers but also on moist rock outcrops, in open woodland and on woodland rides. Lowland. GB 1043+94*, IR 99+11*.

Dioecious. Sex organs tend to develop after mild winters; severe frost damages the thalli and receptacles then fail to develop. Male plants are less frequent than females. Sporophytes are very rare and effective reproduction is by gemmae, which are almost always present.

Widespread in Europe. The native range of *Lunularia*, like that of many synanthropic weeds, cannot be delimited, but it was probably centred on the Mediterranean.

Lunularia could possibly be native in England; it has been found amongst bryophyte material used to plug the seams of a Bronze Age canoe discovered in Lincolnshire, and it was known to Ray (1686) in Essex. In Scotland it is certainly an established introduction.

C. D. PRESTON

305

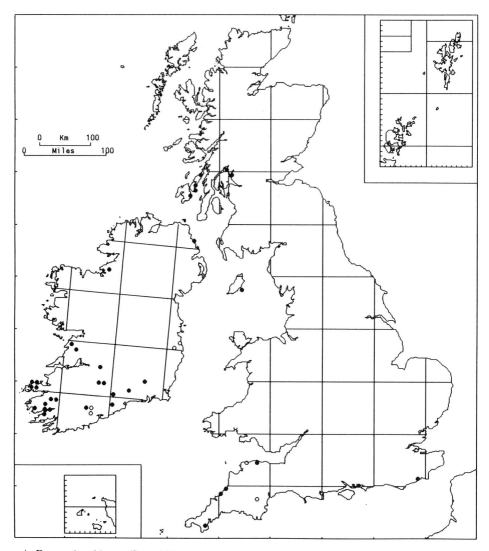

77/1. Dumortiera hirsuta (Swartz) Nees

This species forms extensive patches on well-shaded, continuously moist, and often dripping rocks and steep soil-banks in wooded glens at low elevations. It is commonly associated with *Chrysosplenium oppositifolium*, *Conocephalum conicum*, *Riccardia chamedryfolia*, *Trichocolea tomentella*, *Thamnobryum alopecurum* and, more rarely, *Jubula hutchinsiae*. It also occurs on dripping rocks and in caves and other shaded recesses by waterfalls in wooded ravines, and in moist block-litters up to at least 330 m in S.W. Ireland. In Cornwall and S.E. Ireland it also occurs on shaded riverbanks that are near high-tide level in wooded estuaries. It tends to favour mildly basic substrata. Outside S.W. Ireland it is restricted to low elevations (0–70 m). GB 12+2*, IR 23+4*.

Dioecious or monoecious, with male plants more frequently recorded than females.

Recorded in Europe from France, Spain, Italy, Portugal. Azores, Canary Islands, Madeira; widespread in warm tropical regions in Asia, Africa and S. America. Also Philippines, Hawaii, Tahiti.

The northernmost known world stations for this mainly tropical species are in Britain.

H. J. B. Birks

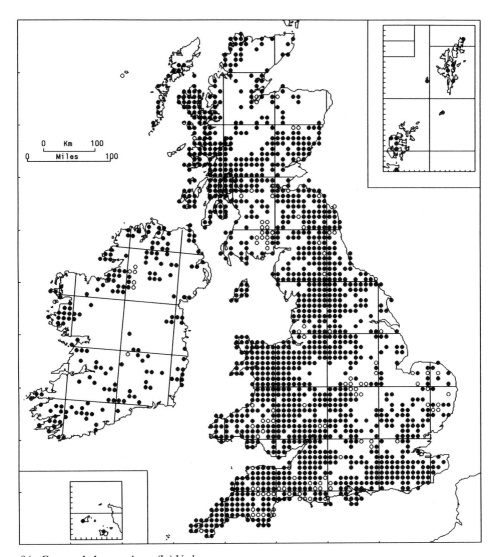

78/1. Conocephalum conicum (L.) Underw.

On neutral or basic substrata in a range of damp, shaded habitats, including stream- and ditch-banks, brick or concrete by streams and rivers, flushes in valley woodland and moist or dripping rocks. It is tolerant of periods of submergence and the thalli, which are strongly bound to the substrate by numerous rhizoids, are not readily dislodged by running water. 0–1000 m (Ben Lawers). GB 1417+114*, IR 178+8*.

Dioecious; capsules occasional, ripe spring. Vegetative propagation is reputedly by tubers on the ventral side of the midrib, but their occurrence in this species needs confirmation.

Circumboreal. Widespread in Europe, but becoming scarcer towards the south and increasingly restricted to the mountains.

C. D. PRESTON

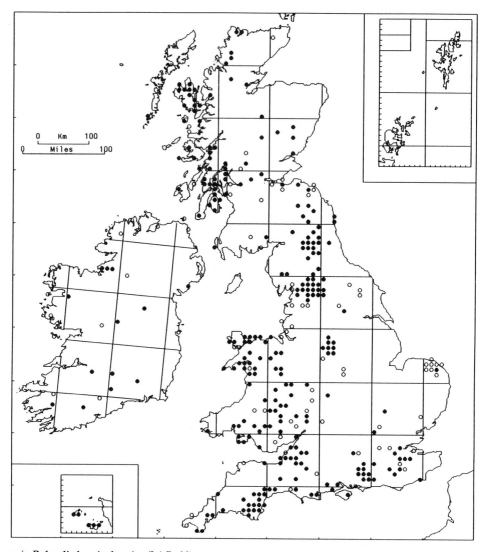

79/1. Reboulia hemisphaerica (L.) Raddi

On sandy lane-sides and hedge-banks, rocky and earthy banks, on rocks, in rock-crevices and on walls, usually in sites which are moist in winter but dry out in summer. Except in S. England it is a distinct calcicole, growing over limestone, schistose or basic igneous rock. 0–460 m (Skye). GB 250+67*, IR 13+12*.

Monoecious or dioecious. Sporophytes occasional, maturing from March to September.

Widespread in Europe, north to S. Scandinavia. It is common in the Mediterranean countries. It has a very wide world distribution, south to Australia and New Zealand.

C. D. Preston

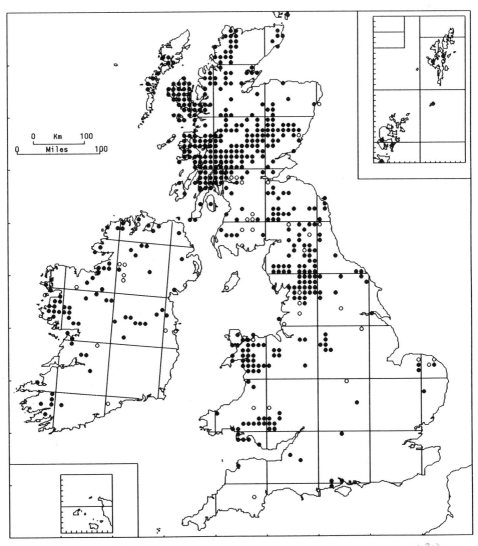

80/1. Preissia quadrata (Scop.) Nees

Found on a wide variety of basic substrates, including soil in calcareous fens and flushes, dune-slacks, stream-banks, moist or dry rocks and cliffs, and on the crumbling mortar of walls. Grows on basalt, gabbro, limestone, sandstone and schistose rocks, especially where they are lightly shaded in gullies, under overhangs or in earthy crevices. A frequent plant in calcareous uplands. 0–1175 m (Ben Lawers). GB 447+34*, IR 73+12*.

Normally dioecious, sometimes monoecious; sporophytes occasional, maturing from May to September. Circumboreal. Europe from Spitsbergen southwards; rare in the Mediterranean.

C. D. PRESTON

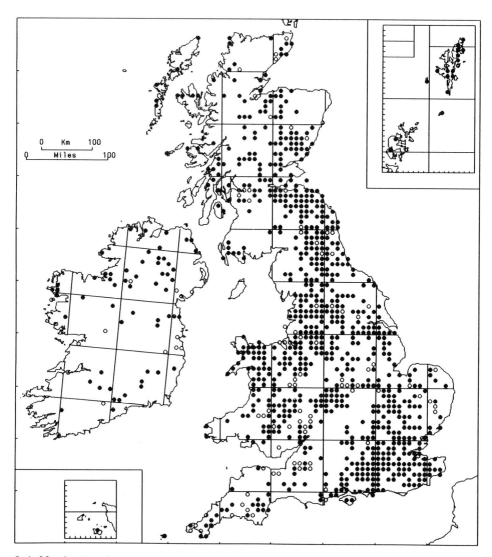

81/1. Marchantia polymorpha L.

Grows on a wide variety of nutrient-rich substrates, in both artificial and natural habitats. Typically found as a weed of plant-pots, greenhouses, gardens and nurseries, on railway-clinker, walls by water and damp wall-bases elsewhere, burnt ground (including bonfire sites), the banks of streams and rivers and in marshes, fens, lakeside swamps and montane springs and flushes. 0–1000 m (Ben Lawers). GB 793+85*, IR 57+13*.

Dioecious. Sex organs and sporophytes, frequent in at least some areas of S.E. England but rare elsewhere, are stimulated by long day-length at fairly high temperatures and hence occur between April and September. Gemmae (in splash-cups) almost always present.

A cosmopolitan species, found throughout Europe and in all continents except Antarctica.

M. polymorpha is a variable plant, here regarded as including the segregate *M. alpestris* (Nees) Burgeff. As a troublesome weed of nurseries it is perhaps the only British hepatic to have any economic significance.

C. D. PRESTON

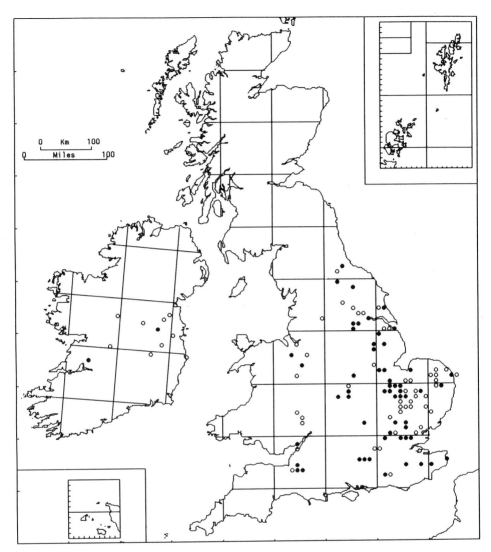

82/1. Ricciocarpos natans (L.) Corda

An aquatic, found floating as scattered or densely crowded plants on the surface of base-rich, eutrophic water in pools, lakes, flooded clay- and gravel-pits, canals, fenland drains and ditches and the sheltered sides of streams and rivers. It is also found as a terrestrial form on damp mud at the water's edge. Associated species include *Lemna minor*, *L. trisulca* and *Riccia fluitans*. Lowland. GB 56+48*, IR 2+10*.

Sex organs and sporophytes have never been found in the British Isles and are rare throughout the European range of the species. Monoecious in North America. Like many plants of the same life-form, it increases clonally. It is allegedly dispersed by water birds, especially swans.

It has an almost cosmopolitan world distribution. Widespread in Europe, extending north to S. Scandinavia.

Reputedly sporadic in its appearance at some sites, although recent systematic observations at Wicken Fen have shown that there it is present every year.

C. D. Preston

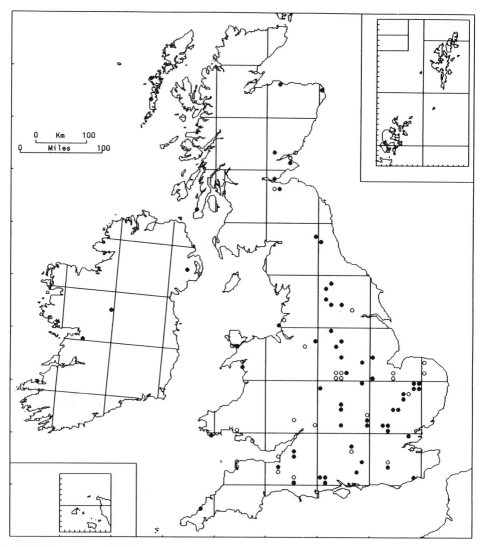

83/1. **Riccia cavernosa** Hoffm.

An annual of seasonally flooded habitats, growing on recently exposed mud at the edge of ponds, lakes, reservoirs and flooded gravel-pits, in hollows in arable fields and on damp sand in dune-slacks. It occurs on a range of substrates from slightly acidic to highly calcareous, often with *Bryum klinggraeffii* and *Physcomitrella patens*. Because of its ecological requirements, *R. cavernosa* is sometimes erratic in its appearance, and field observations suggest that its spores must remain viable for several years. Lowland. GB 61+29*, IR 4.

Monoecious; it fruits abundantly in summer and autumn.

A cosmopolitan species, recorded from all continents except Antarctica. Widespread in Europe, but rare in the Mediterranean region.

R. cavernosa and the related *R. crystallina* have been distinguished in Britain only relatively recently (Paton, 1967a). Earlier records are assumed to refer to *R. cavernosa*, the commoner plant.

C. D. Preston

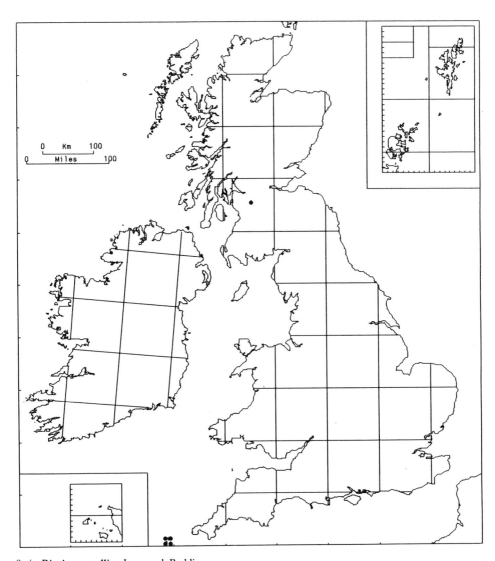

83/2. Riccia crystallina L. emend. Raddi

Locally abundant in bulb- and potato-fields and also occurring on paths in the Isles of Scilly, growing on sandy soil with associates which include *Riccia sorocarpa*, *Sphaerocarpos michelii*, *S. texanus*, *Bryum bicolor*, *B. rubens* and *Pottia truncata*. It has also been recorded (in 1967) from mud on the sides of Waulkmill Glen Reservoir in Renfrewshire, with *Bryum klinggraeffii*, *Leptobryum pyriforme*, *Physcomitrella patens*, *Physcomitrium pyriforme*, *Pottia truncata* and *Pseudephemerum nitidum*. Lowland. GB 5.

Monoecious; fruits abundantly in winter and spring.

A southern species, which tends to replace the closely related *R. cavernosa* in the Mediterranean region and is also widespread in the Southern Hemisphere. It was probably introduced to Scilly with bulbs or other horticultural imports. Its occurrence in Renfrewshire, well to the north of its main range, is most unexpected.

C. D. PRESTON

313

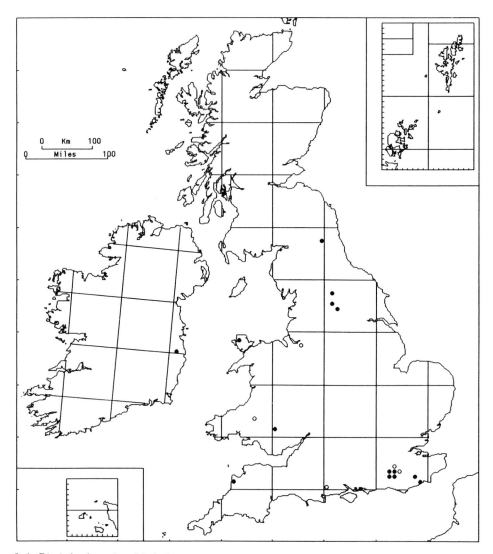

83/3. **Riccia huebeneriana** Lindenb.

A colonist of recently exposed mud at the edge of large ponds, lakes and reservoirs. *R. huebeneriana* is similar in its ecology to the closely related *R. cavernosa* but unlike that species it is distinctly calcifuge. Associated species include *Fossombronia wondraczekii, Riccia glauca, Leptobryum pyriforme, Physcomitrella patens* and *Pseudephemerum nitidum*. 0–450 m (Neuadd Reservoir). GB 14+5*, IR 1.

Monoecious. Sporophytes abundant, maturing in late summer and autumn.

Europe from S. Scandinavia southwards, but not recorded from the eastern Mediterranean. Distribution elsewhere uncertain because of confusion with other species.

C. D. PRESTON

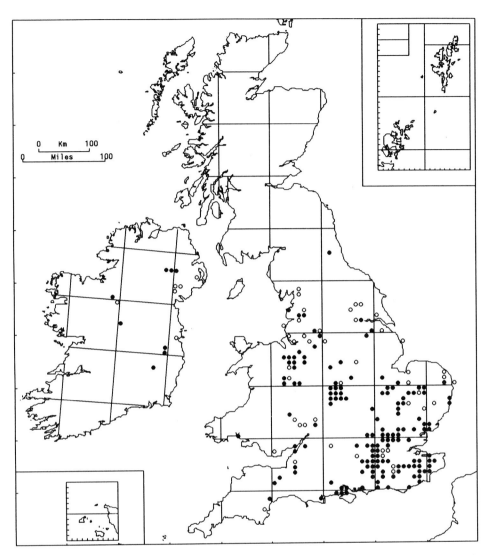

83/4. Riccia fluitans L. emend. Lorbeer

An aquatic species, found as scattered plants or in dense masses floating on the surface of water in ponds, clay- and marl-pits, ditches, canals and in sheltered swamps at the edge of lakes. It occurs in acidic or basic, mesotrophic to eutrophic water over a range of substrates including clay and peat. It is also found terrestrially on mud or on flat, damp stonework by the water's edge. *Lemna minor, L. minuscula* and *L. trisulca*, vascular plants of the same life-form, are characteristic associates. Lowland. GB 144+45*, IR 8+6*.

The species is dioecious but sex organs are rare. Sporophytes have only once been recorded in Britain, in late summer (Paton, 1973b). Reproduction is by vegetative growth. In autumn, plants sink to the bottom of the water, where they overwinter.

Outside Europe it has an almost cosmopolitan distribution. Widespread in Europe, extending north to S. Scandinavia.

C. D. PRESTON

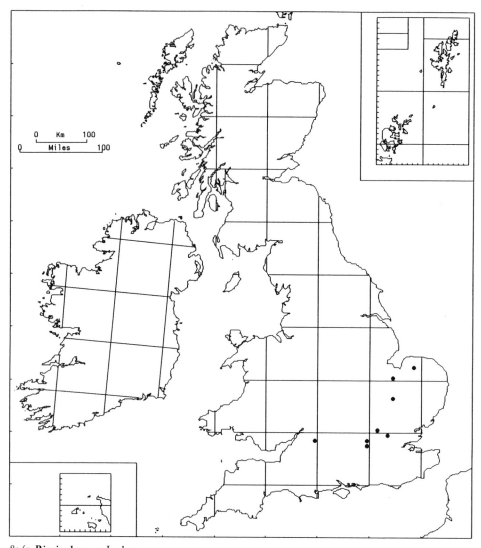

83/5. Riccia rhenana Lorbeer

Occurs as an aquatic floating on the surface of shallow water in ornamental ponds, natural pools and flooded clay- and gravel-pits. It also grows in a modified terrestrial form on mud or decaying vegetation by the water's edge. Lowland. GB 8.

Gametangia and sporophytes unknown, both in the British Isles and elsewhere.

Primarily a C. European species, but extending north to Scandinavia and south to Portugal. S.W. Asia.

Riccia rhenana is grown by aquarists and the British records probably all result from the accidental or deliberate release of cultivated material. It was first recorded in 1952, in an artificial concrete pond in Surrey. It became well established in Madingley Brick Pits, near Cambridge, where it persisted for at least 20 years.

C. D. PRESTON

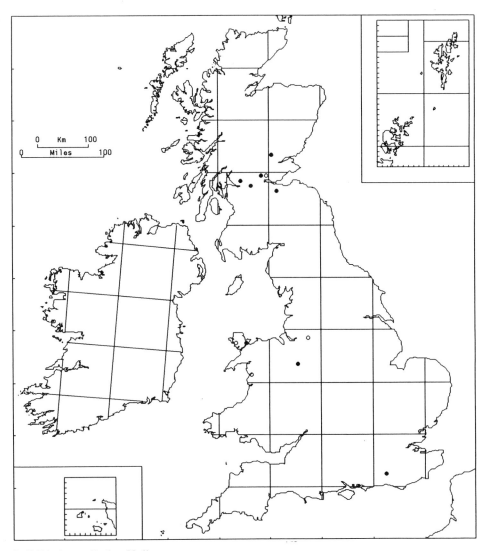

83/6. Riccia canaliculata Hoffm.

A calcifuge ephemeral, confined to recently exposed mud and damp sandy ground at the edge of ponds, lakes and reservoirs. This habitat is present only when water-levels are low and the species is consequently rather erratic in its appearance. It presumably survives periods of high water-level as dormant spores. Lowland. GB 8+4*.

    Monoecious. Sporophytes abundant, recorded from July to February.

    Widespread in Europe. N. Africa, Macaronesia; possibly also N. America.

    R. canaliculata may be under-recorded not only because of its restriction to a rather transient habitat but also because it is superficially similar to other *Riccia* species. Nevertheless, there is little doubt that it is a very local species.

C. D. Preston

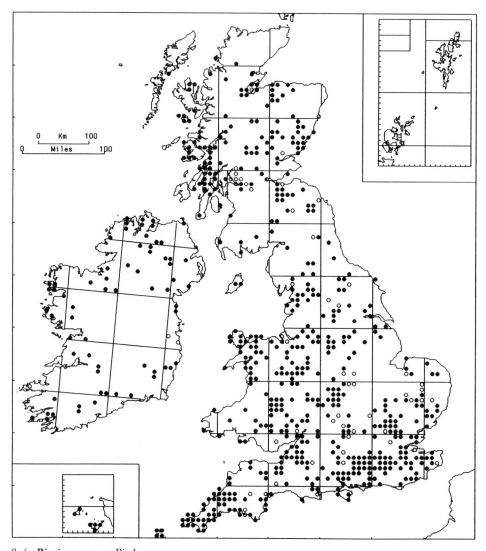

83/7. **Riccia sorocarpa** Bisch.

An ephemeral of sparsely vegetated calcareous or base-poor substrates in sites where the cover of perennial plants is reduced by disturbance or summer drought. It grows in a wide range of such habitats: on thin soil over rocky cliff-slopes, limestone hills and basic igneous rocks; on paths and tracks; in quarries and gravel-pits; on damp clay or gravelly soil in disturbed woodland rides; on earth-topped walls; in stubble-fields, potato-fields, Cornish flower-fields and as a weed in gardens. It is occasionally found on open base-rich cliff-ledges in montane areas. In non-calcareous habitats it is often accompanied by *R. glauca*. 0–800 m (Skye). GB 526+53*, IR 64+4*.

Monoecious. Sporophytes abundant, recorded in almost all months but most frequent in spring and autumn.

Widespread in Europe although absent from areas of continental climate in the north. A cosmopolitan species, recorded from all continents except Antarctica.

C. D. Preston

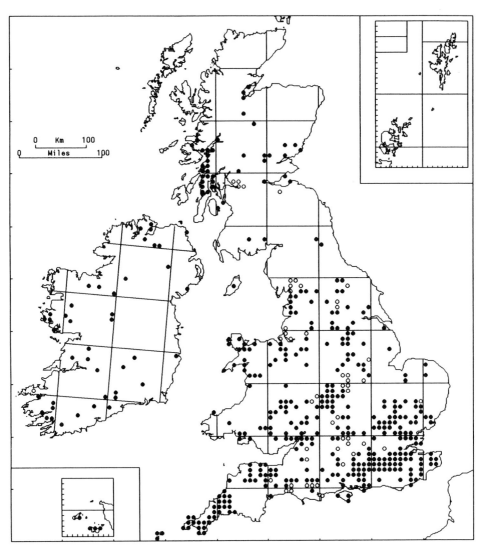

83/8. **Riccia glauca** L.

An ephemeral of disturbed, acidic soil, most characteristically found on base-poor clays and sands in stubble fields and on woodland rides. It is also recorded from a range of other habitats including quarries, gardens, earth-topped walls, riversides, streamsides, lakeside mud, shallow soil on coastal cliff slopes, earthy banks near the sea and (in Cornwall) china-clay workings and flower-fields. Although normally behaving as a calcifuge, it is occasionally found in apparently calcareous habitats such as winter-flooded hollows in arable fields on chalk. Lowland. GB 378+49*, IR 43+1*.

Monoecious. Sporophytes abundant, recorded in almost all months but most frequent in spring and autumn.

Widespread in Europe from S. Scandinavia southwards. N. Africa, Macaronesia, Asia, N. America, Australasia.

C. D. Preston

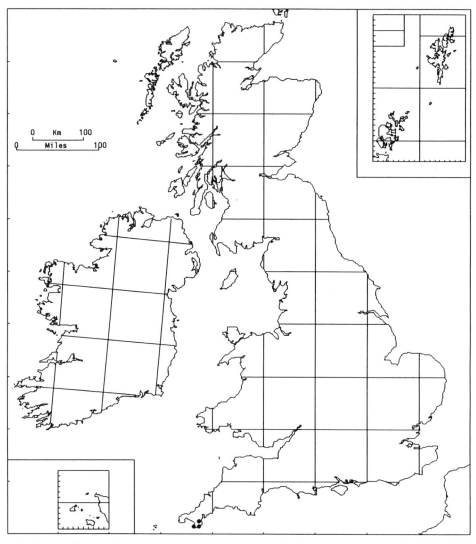

83/9. Riccia bifurca Hoffm.

In partial rosettes or intricate mats in damp or wet hollows, on moist slopes or on tracks across heaths and cliff-tops. It grows on muddy, gravelly or peaty soil, sometimes accompanied by *R. beyrichiana*. It behaves as a perennial in permanently moist places but is apparently annual in exposed, sunny sites. Lowland. GB 3.

Monoecious. Sporophytes frequent, maturing from January to June.

Widespread in Europe from Iceland and Scandinavia south to the Mediterranean. N. America, Australia, New Zealand.

Until recently British bryologists have misunderstood this species. The above account is based on Paton's revision of the relevant material (Paton, 1980).

C. D. PRESTON

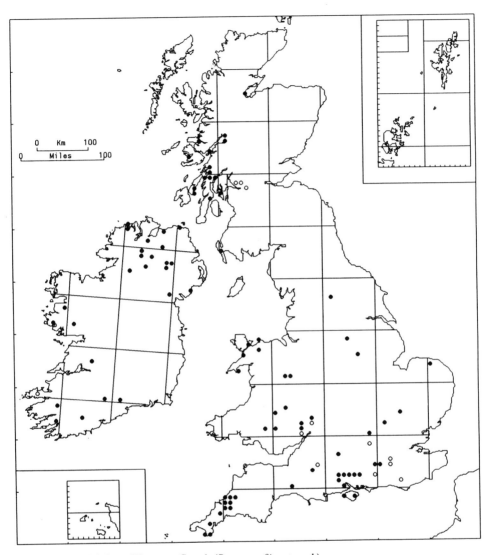

83/10. **Riccia subbifurca** Warnst. ex Crozals (*R. warnstorfii* auct. angl.)

On moist, thinly vegetated soil on rock-ledges and rocky slopes, lane-, stream- and ditch-banks, tracks through woodland and forestry plantations, sandy riversides, the edges of pools and reservoirs and in gardens and arable fields. It is usually found on acidic soils but is also recorded over chalk, limestone and basic igneous rocks. *R. sorocarpa*, which grows in a similar range of habitats, is a frequent associate. Mainly lowland but to 800 m (Creag an Lochain). GB 62+13*, IR 25+1*.

Monoecious; sporophytes frequent, recorded in most months.

W. and S. Europe. Africa, Macaronesia.

Riccia warnstorfii Limpr. has recently been revised (Paton, 1990b). Material from nearly 80 different sites in the British Isles was found to be referable to *R. subbifurca* and no *R. warnstorfii* was seen. All records of *R. warnstorfii* have been transferred to *R. subbifurca*.

C. D. PRESTON

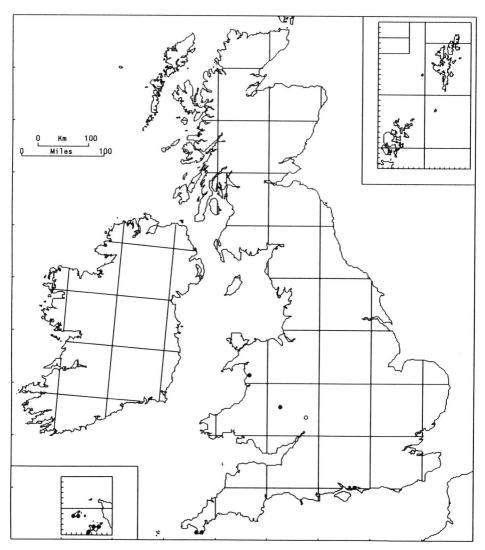

83/11. **Riccia nigrella** DC.

On shallow soil on cliff ledges, rocky slopes, wall-tops, banks and paths, in sites which are moist in winter but regularly desiccated by summer drought and where competition from robust perennial plants is therefore low. In Guernsey it grows on species-rich ledges on S.-facing granite cliffs, where its associates include *Juncus capitatus*, *Lotus hispidus*, *Mibora minima*, *Poa infirma*, *Polycarpon tetraphyllum*, *Romulea columnae*, *Scilla autumnalis* and *Fossombronia husnotii*, all annuals or geophytes near the northern edge of their range. It was recorded from fields at Redmarley D'Abitot (between Gloucester and Malvern) in 1912 but has never been rediscovered there. Lowland. GB 4+1*.

Monoecious. Sporophytes frequent, maturing from December to May.

A Mediterranean-Atlantic species which reaches its northern limit in the British Isles. N. Africa, Macaronesia, S.W. Asia, N. America, Australia.

C. D. PRESTON

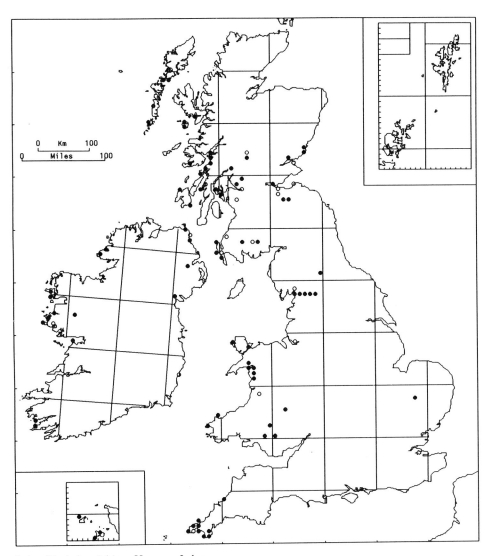

83/12. Riccia beyrichiana Hampe ex Lehm.

A calcifuge, growing on soil which is damp, flushed or flooded in winter. It occurs on tracks, paths, banks and rocky slopes, often near the sea, on the floors of abandoned quarries, in moist hollows in dunes or sandy heathland and on mud at the edge of reservoirs. There are a few records from peat and soil over limestone rocks. Rarely found in cultivated fields. Mainly lowland but to 500 m in N. Wales (Moel Hebog) and 600 m in Scotland (Creag an Lochain). GB 63+15*, IR 15+3*.

Monoecious; sporophytes frequent, maturing in all months. The spores are long-lived and in dry storage have been shown to retain viability for 30 years.

W. and S. Europe, extending north to Iceland and S. Norway and east to Yugoslavia. N. Africa, Azores, N. America, Greenland.

C. D. PRESTON

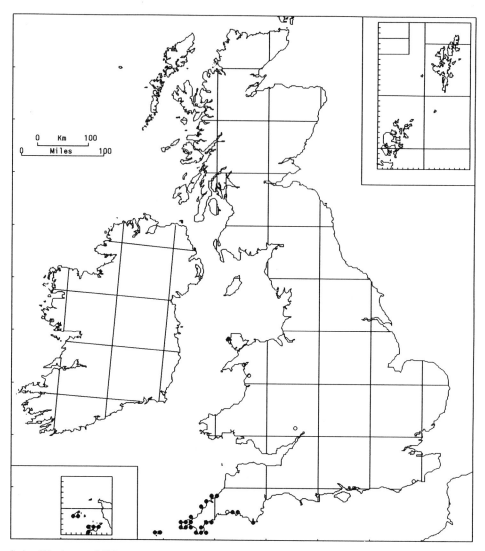

83/13. Riccia crozalsii Lev.

Grows on shallow soil over cliff-slopes, sandy or rocky banks and on coastal paths, in habitats where competition from more robust perennial species is reduced by summer drought. Lowland. GB 23+5*.

Monoecious; sporophytes frequent, maturing from February to June.

One of the more widespread *Riccia* species in the Mediterranean, extending northwards along the Atlantic coast to the British Isles. N. Africa, Macaronesia, S.W. Asia, Australia.

This taxon has, at times, been confused with *R. subbifurca*. Paton (1980) recently revised the relevant material.

C. D. Preston

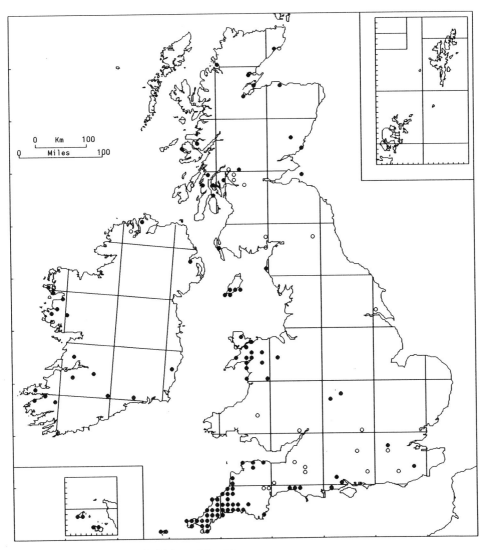

84/1. **Anthoceros punctatus** L. (*A. husnotii* Steph.)

A colonist of fine-textured, wet loamy and clayey soils in fallow fields, cart-tracks, sites of erosion by streams and rivers and, less commonly, wet rocks. Soil pH ranges from slightly alkaline to mildly acidic. Lowland. GB 92+25*, IR 18+3*.

Monoecious; sporophytes maturing throughout the year in persistently wet sites. Without special means of asexual reproduction.

A Mediterranean-Atlantic species, becoming rarer in C. Europe. Occurring widely outside Europe in northern and southern Africa, Macaronesia, N. and S. America, Asia.

M. E. NEWTON

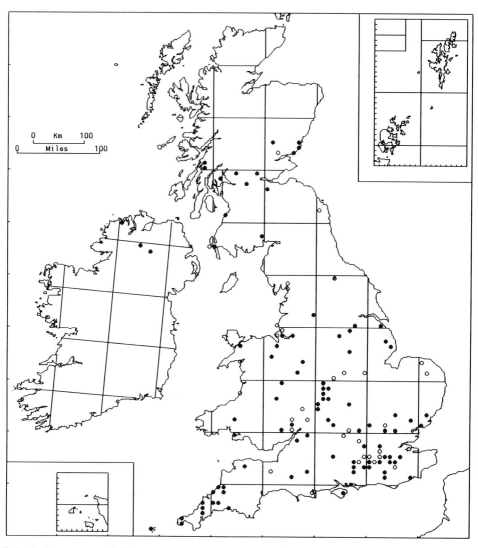

84/2. Anthoceros agrestis Paton

A summer annual on neutral or slightly alkaline, water-retentive disturbed clay and loam, especially in arable fields. It has been recorded only in lowland areas to an altitude of about 170 m (Paton, 1979). GB 92+32*, IR 3.

Monoecious; sporophytes maturing July to December. Without special means of asexual reproduction.

Outside Britain, this species occurs in N., E. and C. Europe. Canaries, Madeira, N. Africa, N. America.

M. E. NEWTON

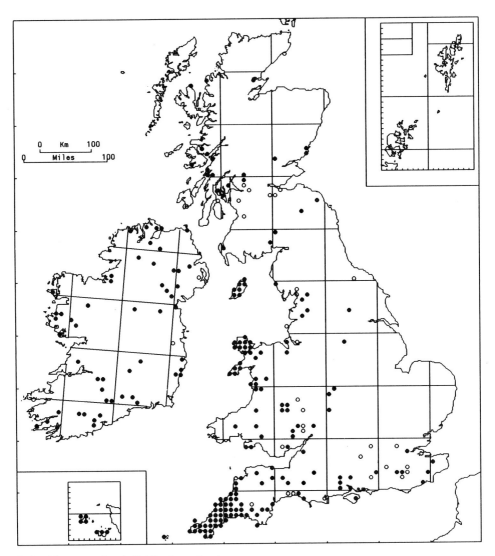

85/1a. Phaeoceros laevis (L.) Prosk. ssp. **laevis**

Commonly on more or less permanently wet, often dripping, clay banks of ditches and streams, less frequently in fallow fields and on muddy tracks. It also occurs occasionally on soft, wet sandstone. Lowland. GB 158+36*, IR 53+3*.

Dioecious; sporophytes are frequent, maturing throughout the year in permanently wet sites. The spores are unusually long-lived, and have germinated after 13 years' dry storage (Proskauer, 1958). Production of perennating tubers is a regular feature, serving as a means of asexual reproduction (cf. Ligrone & Lopes, 1989).

This subspecies is known to occur outside Britain in S. and W. Europe. Its distribution outside Europe is still imperfectly known.

Few recorders have distinguished the two subspecies of *P. laevis*, but Paton (1973c) considered ssp. *laevis* to be the commoner in Britain. The map therefore includes all records except those confirmed as ssp. *carolinianus*, although this may overemphasize the presence of ssp. *laevis*.

M. E. NEWTON

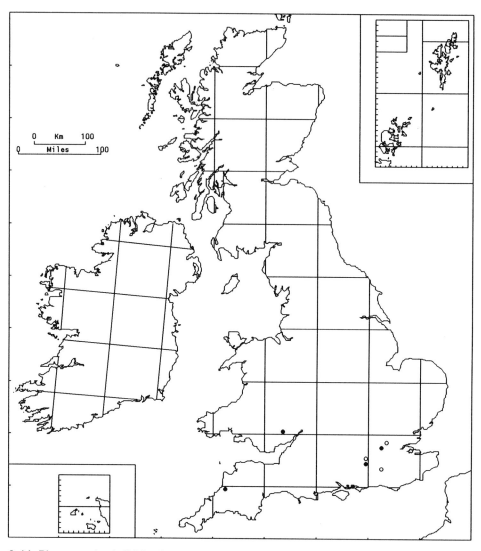

85/1b. **Phaeoceros laevis** (L.) Prosk. ssp. **carolinianus** (Michx.) Prosk.

Occurring as an annual in temporary habitats afforded by moist or wet soil in arable fields and by ditches. Lowland. GB 4+3*.

Monoecious; sporophytes maturing September to December. Tubers have not been reported for this subspecies in Britain, although they are known to occur elsewhere.

This is the commoner, perhaps the only, subspecies in C. Europe, extending north to S. Scandinavia and Iceland; rare or absent in the Mediterranean region. It has a wide world distribution.

In view of the relative frequency and distribution of the two subspecies in Europe, Paton (1973c) considers that ssp. *carolinianus* will be found to extend beyond the southern counties in Britain.

M. E. NEWTON

MAPS OF ENVIRONMENTAL FACTORS

The British Isles are highly diverse, both climatically and topographically. The bryophyte flora reflects this diversity. Eight of the more important environmental factors are mapped here, of which four are climatic, one topographic, one geological and one edaphic; the remaining environmental factor is a measure of atmospheric pollution.

Climatic data were digitized from transparent map overlays already published by the Institute of Terrestrial Ecology (1978). Maps were then processed on a SysScan mapping workstation and plotted with a high-resolution pen plotter.

The other environmental maps have been produced by the same procedure as the maps of species distribution, using data values on a 10-km grid. Data for three of the maps (maximum altitude, bog peat and calcareous rocks) were taken from an ITE database of land characteristics (Ball, Radford & Williams, 1983) for Great Britain. For Ireland, maximum altitude in each 10-km square was read from Irish Ordnance Survey maps, the incidence of bog peat was taken from the *Atlas of Ireland* (Irish National Committee for Geography, 1979), and the occurrence of calcareous rocks was transferred from the overlays in Perring & Walters (1962), converted to the Irish grid. The data for bog peat were supplemented by maps due to Taylor (1983), especially for Wales. Calcareous fen peats do not have a distinctive bryophyte flora and are not mapped.

The final map, of atmospheric sulphur dioxide, was prepared from data kindly supplied in machine-readable medium by Warren Spring Laboratory.

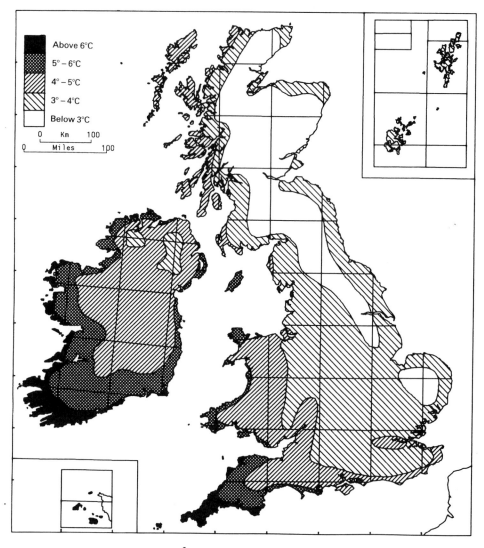

January mean temperature

Values are corrected to sea level. Isopleths are spaced at intervals of 1°C, except for the extreme south-west of Ireland and England, where small areas have a January mean in excess of 7°C.

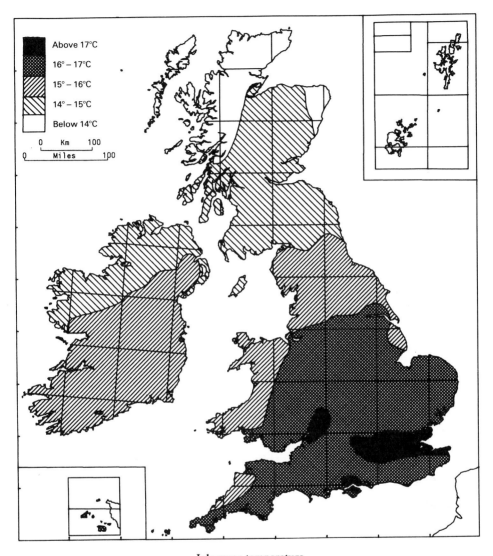

July mean temperature

Values are corrected to sea level. Isopleths are spaced at intervals of 1°C.

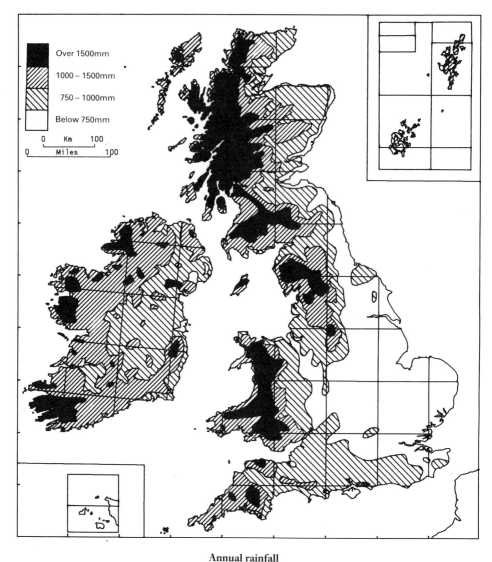

Annual rainfall

The map shows total precipitation, of which a small amount falls as snow, especially in the higher Scottish mountains.

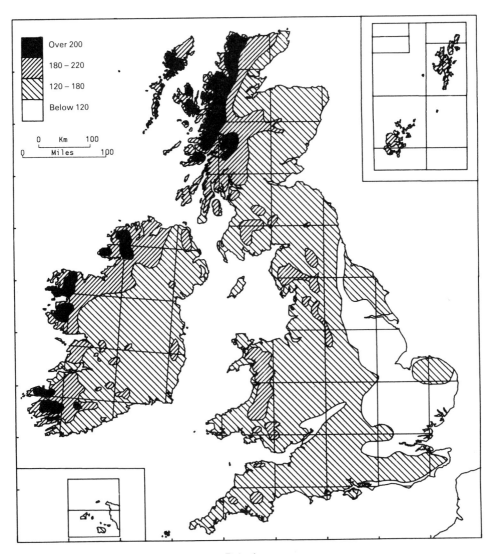

Rain days

For the purpose of the map, a rain day is defined as a day with at least 1 mm of precipitation.

333

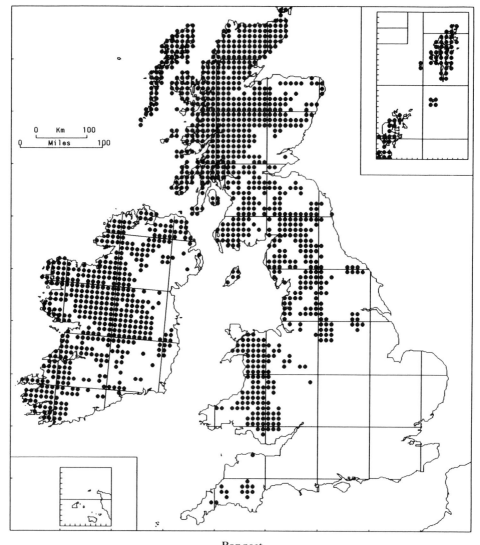

Bog peat

The map indicates grid squares with more than 1 per cent of their land area covered by peat to at least 50 cm depth. Bog peat is acid peat, which could support sphagnum growth.

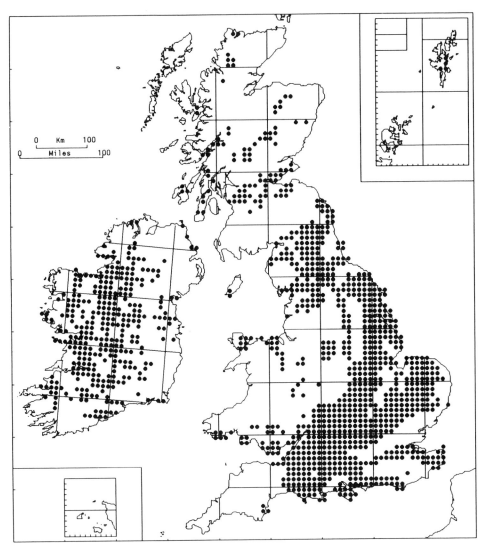

Calcareous rocks

The map indicates grid squares with chalk, limestone or metamorphic calcareous rock underlying at least 5 per cent of the land area of the square.

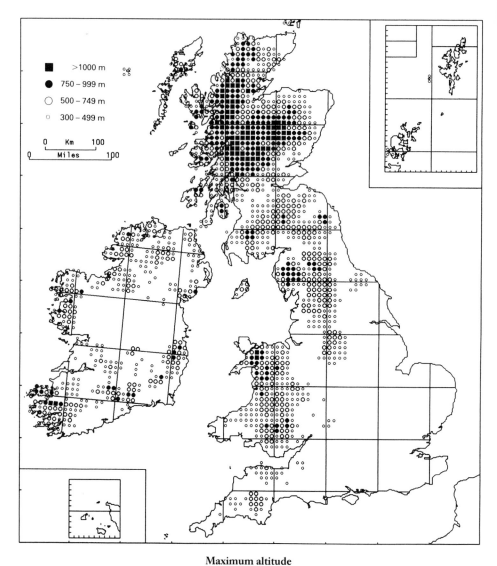

Maximum altitude

10-km squares of the National Grid are classified according to the maximum altitude of land occurring in that square.

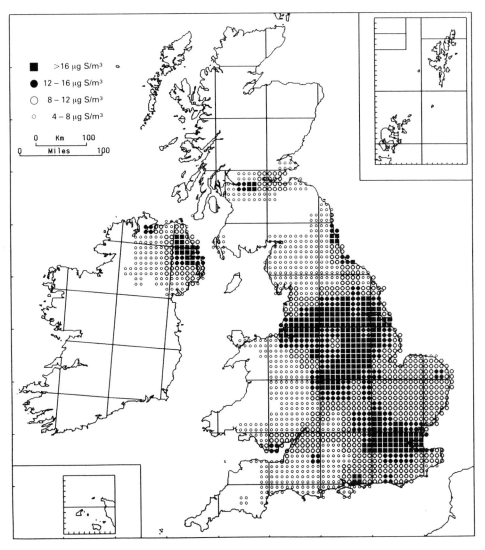

Sulphur dioxide

The map shows the estimated mean annual concentration of sulphur dioxide in the atmosphere in 1987, expressed as the mass of sulphur per cubic metre. Only the United Kingdom has been mapped; comparable data were not available for the Irish Republic. Estimates were made by Warren Spring Laboratory on the basis of a mathematical model, which relies on a rather limited number of measurement stations (mainly urban) together with information on emissions from factories and other sources. Emissions of sulphur dioxide have fallen sharply during the 30 years of the BBS Mapping Scheme; in 1960, atmospheric concentrations would have been much higher.

BIBLIOGRAPHY

Titles of periodicals are abbreviated according to the *World List of Scientific Periodicals* (Brown & Stratton, 1963–65).

Allen, D. E., 1976. *The Naturalist in Britain*. London.

——, 1981. Sources of error in local lists. *Watsonia* **13**, 215–220.

Appleyard, J., 1970. A bryophyte flora of North Somerset. *Trans. Br. bryol. Soc.* **6**, 1–40.

Bagnall, J. E., 1896. The mosses and hepatics of Staffordshire. *J. Bot., Lond.* **34**, 72–77, 108–114.

Ball, D. F., Radford, G. L. & Williams, W. M., 1983. *A Land Characteristic Data Bank for Great Britain*. (Bangor Occasional Paper No. 13). Bangor.

Berkeley, M. J., 1863. *Handbook of British Mosses*. London.

Bisang, I., Geissler, P. & Schumacker, R., 1986. *Harpalejeunea ovata* (Spruce) Schiffn., *Plagiochila exigua* (Tayl.) Tayl. et *Frullania jackii* Gott. à Madonna del Sasso (Tessin, Suisse) et leur repartition européenne. *Candollea* **41**, 413–422.

Boswell, H., 1886. Mosses and Hepaticae. *In:* G.C. Druce, *The Flora of Oxfordshire*, pp. 403–433. Oxford.

Braithwaite, R., 1880. *The Sphagnaceae or Peat-mosses of Europe and North America*. London.

——, 1887–1905. *The British Moss-Flora*. 3 vols. London.

Brown, P. & Stratton, G. B. (eds), 1963–65. *World List of Scientific Periodicals*, edn 4. 3 vols. London.

Carrington, B., 1874–75. *British Hepaticae*. London.

Colgan, N. & Scully, R. W., 1898. *Contributions towards a Cybele Hibernica*, edn 2. Dublin.

Cooke, M. C. [1865]. *Easy Guide to the Study of British Hepaticae*. London.

——, 1894. *Handbook of British Hepaticae*. London.

Corley, M. F .V. & Hill, M. O., 1981. *Distribution of Bryophytes in the British Isles: A Census Catalogue of their Occurrence in Vice-counties*. Cardiff.

Davies, H., 1813. *Welsh Botanology*. London.

Dickson, J., 1785–1801. *Fasciculus Plantarum Cryptogamicarum Britanniae*. 4 vols. London.

Dillenius, J. J., 1724. *Historia Muscorum*. Oxford.

Dixon, H. N., 1896. *The Student's Handbook of British Mosses*. Eastbourne.

——, 1899. The moss flora of Northamptonshire. *J. Northampt. nat. Hist. Soc.* **10**, 183–190, 217–222, 239–258.

——, 1924. *The Student's Handbook of British Mosses*, edn 3. Eastbourne.

Druce, G. C., 1922. The mosses and liverworts of Oxfordshire. *Proc. Rep. Ashmol. nat. Hist. Soc. Oxf.* **1921**, 25–63.

––––––– & Vines, S. H., 1907. *The Dillenian Herbaria*. Oxford.

Duckett, J. G. & Clymo, R. S., 1988. Regeneration of bog liverworts. *New Phytol.* **110**, 119–127.

Duncan, J. B., 1926. *A Census Catalogue of British Mosses*, edn 2. Berwick-upon-Tweed.

–––––––, 1935. *Census Catalogue of British Mosses*, edn 2, supplement. Berwick-upon-Tweed.

Duncan, U. K., 1966. A bryophyte flora of Angus. *Trans. Br. bryol. Soc.* **5**, 1–82.

Fletcher, H. R. & Brown, W. H., 1970. *The Royal Botanic Garden Edinburgh 1670–1970*. Edinburgh.

Foster, W. D., 1979. The history of the Moss Exchange Club. *Bull. Br. bryol. Soc.* **33**, 19–26.

Gardiner, J. C., 1981. A bryophyte flora of Surrey. *J. Bryol.* **11**, 747–841.

Gardiner, W., 1846. *Twenty Lessons on British Mosses*. Dundee.

–––––––, 1848. *The Flora of Forfarshire*. London.

Greville, R. K., 1822–28. *Scottish Cryptogamic Flora*. 6 vols. Edinburgh.

Grolle, R., 1964. *Jamesoniella carringtonii* – eine *Plagiochila* in Nepal mit Perianth. *Trans. Br. bryol. Soc.* **4**, 653–663.

–––––––, 1969. Die Verbreitung von *Pedinophyllum* in Europa. *Herzogia* **1**, 105–110.

––––––– & Schumacker, R., 1982. Zur Synonymik und Verbreitung von *Plagiochila spinulosa* (Dicks.) Dum. und *P. killarniensis* Pears. *J. Bryol.* **12**, 215–225.

Grubb, P. J., 1970. Observations on the structure and biology of *Haplomitrium* and *Takakia*, hepatics with roots. *New Phytol.* **69**, 303–326.

Hardy, J., 1868. The moss flora of the Eastern Borders. *Hist. Berwicksh. Nat. Club* **5**, 443–475.

Hedwig, J., 1801. *Species Muscorum Frondosorum*. Leipzig.

Hobkirk, C. P., 1873. *A Synopsis of the British Mosses*. London.

Hooker, W. J., 1812–16. *British Jungermanniae*. London.

–––––––, 1821. *Flora Scotica*. London.

–––––––, 1833. *The English Flora of Sir James Edward Smith. Class XXIV. Cryptogamia*. London.

––––––– & Taylor, T., 1818. *Muscologia Britannica*. London.

––––––– & –––––––, 1827. *Muscologia Britannica*, edn 2. London.

Horrell, E. C., 1898. The distribution of British mosses. *J. Bot., Lond.* **36**, 60–62.

Hudson, W., 1762. *Flora Anglica*. London.

Ingham, W. (ed.), 1907. *A Census Catalogue of British Mosses*. York.

–––––––, 1913. *A Census Catalogue of British Hepatics*, edn 2. Darwen.

Institute of Terrestrial Ecology, 1978. *Overlays of Environmental and other Factors for use with Biological Records Centre Distribution Maps*. Cambridge.

Irish National Committee for Geography, 1979. *Atlas of Ireland*. Dublin.

Isoviita, P., 1970. Dillenius's 'Historia Muscorum' as the basis of hepatic nomenclature, and S. O. Lindberg's collection of Dillenian bryophytes. *Acta bot. fenn.* **89**, 1–28.

Jameson, H. G., [1893]. *Illustrated Guide to British Mosses*. Eastbourne.

Jones, E. W., 1952. A bryophyte flora of Berkshire and Oxfordshire. I. Hepaticae and Sphagna. *Trans. Br. bryol. Soc.* **2**, 19–50.

———, 1953. A bryophyte flora of Berkshire and Oxfordshire. II. Musci. *Trans. Br. bryol. Soc.* **2**, 220–277.

———, 1957. *Lophozia opacifolia* Culmann in Scotland. *Trans Br. bryol. Soc.* **3**, 180.

———, 1958. An annotated list of British hepatics. *Trans. Br. bryol. Soc.* **3**, 353–374.

———, 1964. African Hepatics XVII. *Gongylanthus* in Tropical Africa. *Trans. Br. bryol. Soc.* **4**, 649–652.

——— & Harrington, A. J., 1983. The hepatics of Sierra Leone and Ghana. *Bull. Br. Mus. nat. Hist. (Bot.)* **11**, 215–289.

Lees, F. A., 1881. *The London Catalogue of British Mosses and Hepatics*, edn 2. London.

———, 1888. *The Flora of West Yorkshire*. London.

Lett, H. W., 1902. *A List, with Descriptive Notes, of all the Species of Hepatics hitherto found in the British Islands*. Eastbourne.

———, 1915. Census report on the mosses of Ireland. *Proc. R. Ir. Acad.* B **32**, 65–166.

Lightfoot, J., 1777. *Flora Scotica*. 2 vols. London.

Ligrone, R. & Lopes, C., 1989. Ultrastructure, development and cytochemistry of storage cells in the 'tubers' of *Phaeoceros laevis* Prosk. (Anthocerotophyta). *New Phytol.* **112**, 317–325.

Linnaeus, C., 1753. *Species Plantarum*. Stockholm.

Little, E. R. B., 1968. The oil bodies of the genus *Riccardia* Gray. *Trans. Br. bryol. Soc.* **5**, 536–540.

Lockhart, N. D., 1989. *Leiocolea rutheana* (Limpr.) K. Müll. new to Ireland. *J. Bryol.* **15**, 525–529.

Long, D. G., 1978. On the distinction between *Scapania aequiloba* and *S. aspera*. *Bull. Br. bryol. Soc.* **31**, 26–29.

———, 1982. *Lophocolea semiteres* (Lehm.) Mitt. established in Argyll, Scotland. *J. Bryol.* **12**, 113–115.

———, Paton, J. A. & Rothero, G. P., 1990. *Marsupella arctica* (Berggr.) Bryhn & Kaal. in Scotland, new to the British Isles. *J. Bryol.* **16**: 163–171.

McArdle, D., 1904. A list of Irish Hepaticae. *Proc. R. Ir. Acad.* B **24**, 387–502.

Mackay, J. T., 1836. *Flora Hibernica. Part Second, comprising the Musci, Hepaticae and Lichenes* [by T. Taylor]. Dublin.

McVean, D. N. & Ratcliffe, D. A., 1962. *Plant Communities of the Scottish Highlands*. London.

Macvicar, S. M. 1904. Census of Scottish Hepaticae. *Ann. Scot. nat. Hist.* **1904**, 43–52.

———, 1905. *Census Catalogue of British Hepatics*. York.

———, 1910. The distribution of Hepaticae in Scotland. *Trans. Proc. bot. Soc. Edinb.* **25**, 1–336.

———, 1912. *The Student's Handbook of British Hepatics*. Eastbourne.

———, 1926. *The Student's Handbook of British Hepatics*, edn 2. Eastbourne.

Malloch, A. J. C., 1972. A note on the ecology of *Gongylanthus ericetorum* in the Lizard peninsula, West Cornwall. *J. Bryol.* **7**, 81–85.

Moore, D., 1876. Report on Irish Hepaticae. *Proc. R. Ir. Acad.*, Ser. 2, **12**, 591–672.

Müller, K., 1954–57. *Die Lebermoose Europas*. 2 vols. Leipzig.

Newton, M. E., 1986. *Pellia borealis* Lorbeer: its cytological status and discovery in Britain. *J. Bryol.* **14**, 215–230.

———, 1988. Cytological diversity in *Pellia endiviifolia* (Dicks.) Dum. *J. Bryol.* **15**, 303–314.

Nicholson, W. E. 1908. The mosses of Sussex. *Hastings E. Suss. Nat.* **1**, 79–110.

———, 1911. The hepatics of Sussex. *Hastings E. Suss. Nat.* **1**, 243–92.

Paton, J. A., 1954. A bryophyte flora of the sandstone rocks of Kent and Sussex. *Trans. Br. bryol. Soc.* **2**, 349–374.

———, 1961. A bryophyte flora of South Hants. *Trans. Br. bryol. Soc.* **4**, 1–83.

———, 1965a. *Census Catalogue of British Hepatics*, edn 4. British Bryological Society.

———, 1965b. *Lophocolea semiteres* (Lehm.) Mitt. and *Telaranea murphyae* sp. nov. established on Tresco. *Trans. Br. bryol. Soc.* **4**, 775–779.

———, 1966. Distribution maps of bryophytes in Britain. *Eremonotus myriocarpus* (Carringt.) Pears. *Trans. Br. bryol. Soc.* **5**, 155.

———, 1967a. *Riccia crystallina* L. and *Riccia cavernosa* Hoffm. in Britain. *Trans. Br. bryol. Soc.* **5**, 222–225.

———, 1967b. *Leptoscyphus cuneifolius* (Hook.) Mitt. with perianths in Ireland. *Trans. Br. bryol. Soc.* **5**, 232–236.

———, 1971. *Southbya tophacea* Spruce in Anglesey. *Trans. Br. bryol. Soc.* **6**, 328–330.

———, 1973a. Taxonomic studies in the genus *Fossombronia* Raddi. *J. Bryol.* **7**, 243–252.

———, 1973b. *Riccia fluitans* L. with sporophytes. *J. Bryol.* **7**, 253–259.

———, 1973c. *Phaeoceros laevis* (L.) Prosk. subsp. *carolinianus* (Michaux) Prosk. in Britain. *J. Bryol.* **7**, 541–543.

———, 1974a. *Fossombronia fimbriata* sp. nov. *J. Bryol.* **8**, 1–4.

———, 1974b. *Lophocolea bispinosa* (Hook. f. & Tayl.) Gottsche, Lindenb. & Nees established in the Isles of Scilly. *J. Bryol.* **8**, 191–196.

———, 1977a. *Metzgeria temperata* Kuwah. in the British Isles, and *M. fruticulosa* (Dicks.) Evans with sporophytes. *J. Bryol.* **9**, 441–449.

———, 1977b. *Plagiochila killarniensis* Pears. in the British Isles. *J. Bryol.* **9**, 451–459.

———, 1979. *Anthoceros agrestis*, a new name for *A. punctatus* var. *cavernosus sensu* Prosk. 1958, *non* (Nees) Gottsche *et al.. J. Bryol.* **10**, 257–261.

———, 1980. Observations on *Riccia bifurca* Hoffm. and other species of *Riccia* L. in the British Isles. *J. Bryol.* **11**, 1–6.

———, 1984. *Cephaloziella nicholsoni* Douin & Schiffn. distinguished from *C. massalongi* (Spruce) K. Müll. *J. Bryol.* **13**, 1–8.

———, 1987. Bulbils on *Telaranea nematodes* (Gott. ex Aust.) Howe in Ireland. *J. Bryol.* **14**, 792–793.

———, 1990a. *Marsupella profunda* Lindb. in Cornwall, new to the British Isles. *J. Bryol.* **16**, 1–4.

———, 1990b. *Riccia subbifurca* Warnst. ex Crozals in the British Isles. *J. Bryol.* **16**, 5–8.

Pearman, M. A., 1979. British bryophyte floras and check-lists, 1954–78. *J. Bryol.* **10**, 561–573.

Pearson, W. H., 1902. *The Hepaticae of the British Isles.* 2 vols. London.

Perring, F. H., Sell, P. D., Walters, S. M. & Whitehouse, H. L. K., 1964. *A Flora of Cambridgeshire*. Cambridge.

—— & Walters, S. M. (eds), 1962. *Atlas of the British Flora*. London.

Perry, A. R. (ed.), 1983. Reminiscences of some members of the British Bryological Society. *Bull. Br. bryol. Soc.* **42**, 17–45.

Power, T., 1845. *The Botanist's Guide for the county of Cork*. Cork.

Praeger, R. L., 1901. Irish Topographical Botany. *Proc. R. Ir. Acad.* **22**, i–clxxxviii, 1–410.

Proctor, M. C. F., 1956. A bryophyte flora of Cambridgeshire. *Trans. Br. bryol. Soc.* **3**, 1–49.

Proskauer, J., 1958. Studies on Anthocerotales. V. *Phytomorphology* **7**, 113–135.

Purchas, W. H. & Ley, A. (eds), [1889]. *A Flora of Herefordshire*. Hereford.

Ray, J., 1660. *Catalogus Plantarum circa Cantabrigiam nascentium*. Cambridge.

——, 1663. *Appendix ad Catalogum Plantarum circa Cantabrigiam nascentium*. Cambridge.

——, 1686–1704. *Historia Plantarum*. 3 vols. London.

——, 1696. *Synopsis Methodica Stirpium Britannicarum*, edn 2. London.

——, 1724. *Synopsis Methodica Stirpium Britannicarum*, edn 3. London.

Relhan, R., 1785. *Flora Cantabrigiensis*. Cambridge.

Richards, P. W., 1963. *Campylopus introflexus* (Hedw.) Brid. and *C. polytrichoides* De Not. in the British Isles; a preliminary account. *Trans. Br. bryol. Soc.* **4**, 404–417.

——, 1979. A note on the bryological exploration of North Wales. *In:* G. C. S. Clarke & J. G. Duckett (eds), *Bryophyte Systematics*, pp. 1–9. London.

——, 1985. The British Bryological Society 1923–83. *In:* R. E. Longton & A. R. Perry (eds), *British Bryological Society Diamond Jubilee*, pp. 3–10. Cardiff.

—— & Smith, A. J. E., 1975. A progress report on *Campylopus introflexus* (Hedw.) Brid. and *C. polytrichoides* De Not. in Britain and Ireland. *J. Bryol.* **8**, 293–298.

—— & Wallace, E. C., 1950. An annotated list of British mosses. *Trans. Br. bryol. Soc.* **1**, Appendix, i–xxxi.

Robinson, J. F., 1871. Obituary notice of William Wilson, author of the "Bryologia Britannica". *Trans. Proc. bot. Soc. Edinb.* **11**, 170–174.

Rose, F., 1949. A bryophyte flora of Kent, I. *Trans. Br. bryol. Soc.* **1**, 202–210.

Sayre, G., 1983. A Thomas Taylor bibliography. *J. Bryol.* **12**, 461–470.

Schuster, R. M., 1981. Evolution and speciation in *Pellia* with special reference to the *Pellia megistospora* – *endiviifolia* complex (Metzgeriales), I. Taxonomy and distribution. *J. Bryol.* **11**, 411–431.

Scott, G. A. M., 1987. Studies in ancient bryology, I. Introduction and liverworts to 1500 A.D. *J. Bryol.* **14**, 625–634.

——, 1988. Studies in ancient bryology, II. Mosses to 1500 A.D. *J. Bryol.* **15**, 1–15.

Scott, L. I., 1947. Report of the annual meeting, 1946. *Trans. Br. bryol. Soc.* **1**, 47–50.

Sherrin, W. R., 1937. *Census Catalogue of British Sphagna*. Berwick-upon-Tweed.

——, 1946. *Census Catalogue of British Sphagna*, edn 2, revised by A. Thompson. Berwick-upon-Tweed.

Side, A. G., 1970. An atlas of the bryophytes found in Kent. *Trans. Kent Fld Club* **4**, 1–140.

Smith, A. J. E., 1978a. *The Moss Flora of Britain and Ireland*. Cambridge.

————, 1978b. *Provisional Atlas of the Bryophytes of the British Isles*. Huntingdon.

————, 1990. *The Liverworts of Britain and Ireland*. Cambridge.

Smith, J. E., 1804. *Flora Britannica*, 3. London.

Sowerby, J., 1790–1814. *English Botany*. 36 vols. London.

Sowerby, J. de C. (ed.), 1829–66. *Supplement to the English Botany of the late Sir J. E. Smith and Mr. Sowerby*. 5 vols. London.

Stark, R. M., 1854. *A Popular History of British Mosses*. London.

Steel, D. T., 1978. The taxonomy of *Lophocolea bidentata* (L.) Dum. and *L. cuspidata* (Nees) Limpr. *J. Bryol.* **10**, 49–59.

Taylor, F. J., 1954. Bryophyte County Floras. I. The Channel Islands, England and Wales. *Trans. Br. bryol. Soc.* **2**, 446–457.

————, 1955. Bryophyte County Floras. II. Scotland and Ireland. *Trans. Br. bryol. Soc.* **2**, 539–551.

Taylor, J. A., 1983. Peatlands of Great Britain and Ireland. *In:* Gore, A. J. P. (ed.), *Mires: Swamp, Bog, Fen and Moor. Regional Studies* (Ecosystems of the World, vol. 4B), pp. 1–46. Amsterdam.

Tripp, F. E., 1874. *British Mosses*. 2 vols. London.

Turner, D., 1804. *Muscologiae Hibernicae Spicilegium*. Yarmouth.

————, 1835. *Extracts from the Literary and Scientific Correspondence of Richard Richardson, M.D., F.R.S., of Bierley, Yorkshire*. Yarmouth.

Váňa, J., 1988. *Cephalozia* (Dum.) Dum. in Africa, with notes on the genus. *Beih. Nova Hedwigia* **90**, 179–198.

Wallace, E. C., 1963. Distribution maps of bryophytes in Britain. *Ptilidium pulcherrimum* (Weber) Hampe. *Trans. Br. bryol. Soc.* **4**, 513.

————, 1979. *Lophocolea bispinosa* (Hook. f. & Tayl.) Gottsche, Lindenb. & Nees established on the Isle of Colonsay, Scotland. *J. Bryol.* **10**, 576–577.

Warburg, E. F., 1963. *Census Catalogue of British Mosses*, edn 3. British Bryological Society.

Watson, E. V., 1955. *British Mosses and Liverworts*. Cambridge.

————, 1985. The recording activities of the BBS (1923–83) and their impact on advancing knowledge. *In:* R. E. Longton & A. R. Perry (eds), *British Bryological Society Diamond Jubilee*, pp. 17–29. Cardiff.

Watson, H. C., 1843. *The Geographical Distribution of British Plants*, edn 3. Part 1. London.

————, 1873–74. *Topographical Botany*. 2 vols. Thames Ditton.

Webb, D. A., 1955. The Distribution Maps Scheme : a provisional extension to Ireland of the British National Grid. *Proc. bot. Soc. Br. Isl.* **1**, 316–319.

Wheldon, J. A., 1898. The mosses of Cheshire. *J. Bot., Lond.* **36**, 302–311.

———— & Wilson, A., 1907. *The Flora of West Lancashire*. Eastbourne.

Wilson, A., 1930. *A Census Catalogue of British Hepatics*, edn 3. Berwick-upon-Tweed.

————, 1935. *Census Catalogue of British Hepatics*, edn 3, supplement. Berwick-upon-Tweed.

————, 1938. *The Flora of Westmorland*. Arbroath.

Wilson, W., 1855. *Bryologia Britannica*. London.

LIST OF LOCALITIES CITED
IN THE TEXT

Localities mentioned in the text are given below with their co-ordinates. Where possible, the 10-km square or squares are given but for larger areas (e.g. Breadalbane, Pennines) the 100-km squares are given. The accompanying map (p. 346) gives the numerical equivalents of the 100-km square alphabetical codes, used below.

A
Abbey Gardens, Tresco, SV81
Achill Island, F50,60,61,70,71 L69,79
Airlie Castle, NO25
An Caisteal, NN31
Aonach Beag, NN17
Appleby, NY62
Applecross, NG74
Arran, NR,NS
Arrochar, NN20

B
Ballaghbeama Gap, V77
Bantry, V94
Barra, NF60,70, NL69,79
Beinn a'Bhuird, NO09
Beinn Dearg, NH28
Beinn Dorain, NN33
Beinn Eighe, NG95,96
Beinn Heasgarnich, NN43
Beinn Ime, NN20
Ben Alder, NN47
Ben Alder Forest, NN57
Ben Cruachan, NN03
Ben Hope, NC44,45
Ben Lawers, NN64
Ben Lomond, NN30
Ben Lui (= Ben Laoigh), NN22
Ben More Assynt, NC32
Ben Nevis, NN17
Ben Vorlich, Dunbartonshire, NN21
Ben Vorlich, Perthshire, NN61
Ben Wyvis, NH46
Benbulbin, G64,74
Benmore Gardens, NS18
Berwyn Mts, SJ02,03,04,13,14
Bidean nam Bian, NN15
Black Notley, TL72
Bodmin Moor, SX
Borrowdale, NY21
Braemar area, NO19

Braeriach, NN99, NH90
Breadalbane, NN

C
Cader Idris, SH71
Cairn Gorm, NJ00
Cairngorm Mountains, NH,NJ,NN,NO
Cam Creag, NN52
Ceunant Llennyrch, SH63
Chisbury Wood, SU26
Coire an Lochain, NH90
Coire an t-Sneachda, NH90
Coire Ardair, NN48
Coire Cheap, NN47
Colonsay, NR38,39,49, NM40
Conival, NC31
Craven Pennines, SD
Creag an Dail Bheag, NO19
Creag an Lochain, NN54
Creag Meagaidh, NN48
Croydon, TQ36
Cruach Ardrain, NN42

E
Einich Cairn, NN99
Enfield Chase, TQ29

F
Findhorn, River, NH,NJ
Fothringham, NO44

G
Galtee Mts, R81,82,92
Glas Tulaichean, NO07
Glen Affric, NH02,12,22
Glen Clova, NO27,37,36
Glen Coe, NN15
Glen Doll, NO27
Glen Duror, NN05
Glen Feshie, NN89
Glen Lochay, NN43,53

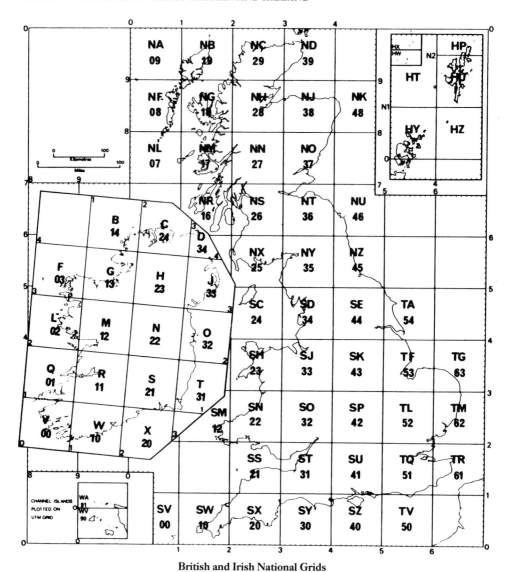

British and Irish National Grids

The map shows the numerical equivalents of the 100 km-square alphabetical codes used in the list of localities cited in the text.

INDEX TO SPECIES IN VOLUME 1